Walter Draper

Gardening in Egypt

A Handbook of gardening for Lower Egypt

Walter Draper

Gardening in Egypt

A Handbook of gardening for Lower Egypt

ISBN/EAN: 9783337069476

Printed in Europe, USA, Canada, Australia, Japan

Cover: Foto ©berggeist007 / pixelio.de

More available books at **www.hansebooks.com**

GARDENING IN EGYPT:

— A —

HANDBOOK OF GARDENING

FOR

LOWER EGYPT.

BY

WALTER DRAPER.

Fellow of the Royal Horticultural Society; late of the Royal Botanic Gardens, Kew; and the Crystal Palace, London; Director of Government Gardens at the Barrage, near Cairo.

LONDON:
L. UPCOTT GILL, 170, STRAND, W.C.

1895.

"It is both the duty and interest of every owner and cultivator of the soil to study the best means of rendering that soil subservient to his own and the general wants of the community; and he who introduces a new and useful *Seed*, *Plant*, or *Shrub* into his district, is a blessing and honour to his country."—SIR J. SINCLAIR.

PRINTED AT THE LONDON AND COUNTY WORKS, DRURY LANE, W.C.

DEDICATED

TO

G. B. ALDERSON, ESQ.,

AS A TOKEN OF RESPECT FOR HIS

MANY ACTS OF KINDNESS TO ME DURING MY

STAY IN RAMLEH.

PREFACE.

So far as I am aware, no book on Egyptian Gardening has yet been published in English. That this little work will fill up the void, I do not for a moment pretend, as the subject is an extensive one, and would require a much larger treatise than the simple handbook I now offer to my readers.

All that I have done is to respond to an invitation to put into a more permanent form the notes that I took for the course of Horticultural lectures, which I delivered last winter in Ramleh.

These notes have been revised and enlarged, and I have added at the end a garden calendar of work for the different months of the year.

I have to thank many of the residents of Alexandria and Ramleh for the kind interest they have shown in my work, and for the practical hints they have given me on Gardening matters.

To Professor Sickenberger, of the School of Medicine, Cairo, I am indebted for information concerning the number of species of the Date Palm, and specimens of new cactaceous and other plants from Abyssinia; also to my friend, J. F. Duthie, Esq., B.A., F.L.S., Director of the Botanical Department, Northern India, for collections of seeds, many of which have grown well and are new to this country; and to the kindness of Dr. Geo. King, C.I.E., LL.D., F.R.S., F.L.S., Director of the Royal Botanic Gardens, Calcutta, by whose help I have been able to introduce several varieties of Orchids and Aroids, that were previously not grown in Egypt.

In describing plants I have endeavoured, when possible, to give the Latin, European, or Native names, which will be found in three distinct types, commencing with the former—for example: **Solanum Melongena**—Egg Plant—*Bydingan*.

RAMLEH,
    EGYPT.
        June, 1895.

CONTENTS.

Chap.		Page
I.—Climate and Soil		1
II.—Laying out a Garden		5
III.—Trees		15
IV.—Palms		29
V.—Shrubs		38
VI.—Climbing Plants		46
VII.—Roses		55
VIII.—Conservatories and Glasshouses		63
IX.—Cactaceous and Succulent Plants		73
X.—Fruit and Vegetables		89
Garden Calendar		105
Index		109

GARDENING IN EGYPT.

CHAPTER I.

CLIMATE AND SOIL.
CLIMATE.

THE close proximity of this country to Europe is often the cause of too little attention being paid to its climatic peculiarities. Egypt is just near enough to the Equator to alter the seasons without entirely changing them, and therefore it is a considerable help to the horticulturist if he makes a careful study of the changes he will have to contend with.

The climate of Egypt is to a great extent influenced by the Nile, but the immense absorbing power of the desert is the country's chief regulator, for, were it not for these large tracts of desert plains, the winter rains of Lower Egypt, would extend far up the Nile Valley, and the great expanse of nearly stagnant water at both mouths of the Nile, would possibly render the fertile Delta, as unhealthy as the pestilent regions of the West Coast of Africa.

The Egyptian year may be divided into two seasons: the hot season from April to November, and the cold season from December to March. February is usually the month for high winds, although occasionally strong winds are experienced even before and after that month; and it is certain that in Alexandria, Ramleh, and other places on the coast, the salt-laden wind from the Mediterranean is undoubtedly the worst enemy to the gardens.

It will therefore behove the judicious horticulturist with an exposed garden in these localities, to see that young trees and tall plants are securely staked, that tender plants are protected, and that all large fan-leaved palms, such as *Latania borbonica*, *Sabal umbraculifera*, and *Washingtonia filifera*, have their leaves securely tied up. The same rule applies to choice tender plants that have stood out in isolated positions during the summer, like *Crotons* and others, which should be taken in at the beginning or middle of November.

During the next three months the thermometer will sometimes fall to 35deg. Fahr. in Alexandria, and 40deg. Fahr. in Cairo; therefore, when there are glass-houses, all available space should be utilised in storing tender plants from the low temperature experienced outside.

The south winds from the desert, known as "Khamseens," which occur in the spring—usually in March and April—are in some cases detrimental to young plants, but the damage done is not nearly so great as that caused to fruit-blossom and tender shoots by the salt winds from the sea. These winds make the plants appear as if they had been burnt, and often strip entire hedges and trees of their foliage. It will be found, however, that where the wind is uninterrupted in its passage, the injury to plants is not so great as if its course is obstructed.

The climate of Cairo, with reference to plant life, varies considerably from that of Alexandria, insomuch that many plants which thrive in one town will not flower, or only partially exist in the other, and then only under the most favourable circumstances. The following instances of this may be quoted: *Grevillea robusta* (Australian Silky Oak), *Caryota urens* (Toddy Palm), and *Bauhinia purpurea*, which all do well in

Climate and Soil.

Cairo, but rarely flower or grow into fine specimens in Alexandria.

It is advisable to remember that plants which have been housed during the winter, cannot at once withstand the full force of the sun, and therefore it is necessary for the first week or so, to place them in a half-shady position in order to subject them to a system of "hardening off," and so prevent their foliage from being scorched by the sun.

At the end of the cold season, about the middle of April, is the time when most of the European plants and annuals are generally at their best. This is also the best season for roses, though some species flower more or less all through the summer, and produce again a second show of bloom in the early autumn.

This influx of flowers of temperate climes, is followed by a still more gorgeous display from the tropical plants. Thus we see to perfection in May, the blue *Jacaranda mimosifolia*, and in June and July, the red *Poinciana regia* of Madagascar, which rank amongst the most brilliant of all Egyptian flowering trees.

SOIL.

In this accommodating Egyptian climate, a sheltered position is perhaps more necessary than the quality of the soil, although the salt sandy soil of Alexandria, has a detrimental effect on all trees and plants, as can easily be seen when the gardens there are compared with those of Cairo.

There are three different kinds of soil, with which one generally comes in contact, when laying out gardens in Egypt:(1st) Either the alluvial Nile mud, which is chiefly met with in Cairo and the neighbourhood. (2nd) The *débris* of Alexandria, which contains a large percentage of salt, sand, and lime. (3rd) Or, again, the sand of Ramleh, from which the place derives its name. But in spite of the drawbacks of a very poor soil, most creditable gardens, producing an abundance of flowers, are seen in Ramleh, and this speaks well of the attention which is paid to them by their owners.

B 2

The advantage inland towns have over those on the coast, from a horticultural point of view, is very great; for not only is the alluvial Nile mud beneficial to the growth of plants, but also the lowness of the land in many places, and the close proximity to the river and canals tends much to create a permanent moist undersoil; consequently the plants being free from salt and lime, either in the soil or in the air, can grow unchecked, and the only danger that is to be feared arises from the cold winds from the desert.

Alexandria, on the other hand, apart from its sandy soil, which refuses to hold the moisture, is in many cases tunnelled by ancient excavations, which often cause the ground to subside; not only therefore is there always a possibility of the rearrangement of an Alexandrian garden becoming necessary, owing to these subsidences, but also the horticulturist has to contend with a peculiar kind of oily dew which is harmful to the plants.

It is thought by some authorities that this oily dew is caused by the infiltration of the sea into the lake, which leaves during its passage a number of poisonous gases, and as the dew does not ascend high enough to become aërated, it is consequently laden when it falls with many of these gases that have risen with it from the soil.

If water is obtainable, it is possible to grow certain plants on any soil, even in the most exposed positions, but it is always advisable to improve the ground by applying decayed garden refuse, Nile mud, and manure. It will also be found to be a great advantage to dig or trench the soil from time to time, so as to allow it to be exposed and become fallow, as the sun in Egypt acts in very much the same way as the frost in Europe in sweetening the ground.

In planting choice specimens—trees, climbers, &c.—it is advisable to add a considerable portion of rich compost in order that the plants may become well established before their roots reach the poorer soil.

CHAPTER II.

LAYING OUT A GARDEN.

A SHELTERED position is the chief thing to be considered in the laying out of a garden as the trees and plants that can be grown, will in a great measure, be governed by the situation and locality in which the garden is made. Ornamentation and style, must be left to the taste and judgment of the owner, for, although we are debarred from growing many of our old English favourites, and the flowers do not have the same smell, yet we have many beautiful tropical and sub-tropical plants, which compensate for their loss, and if it is not possible to have a strictly European garden, it is at least possible to have a very pretty Egyptian one, subject however, to the conditions of the locality.

No one should suppose that he can have a well-kept garden, that will be a continual source of pleasure, unless he is to a considerable extent, his own head gardener, and although scarcely two gardens are alike in style and position, yet there are certain rules in landscape gardening, which hold good abroad, as well as at home.

Thus in laying out ornamental grounds, either public or private, the first thing that is necessary is taste, and a careful observation of nature, which might often be copied with advantage; and the second is a knowledge of the habit and position most suitable for each plant, so that things may grow and increase in beauty, rather than become an unsightly and entangled mass.

One of the most important things in the laying out of a private garden, is the view from the front of the house. Should the position be elevated, as is often the case in Ramleh, the natural views, which add much to the beauty of a place, should be well considered, and the view opened up, rather than shut out; in such cases, trees that are planted should be utilised for shade and shelter, and at the same time, if necessary, they should hide any unsightly object.

Although natural elevations are always desirable, yet it is quite possible to create a very pleasing effect with grounds that are flat. In gardens of the latter kind, surrounded by desert, or uninteresting views, the chief thing to be considered is a background of trees, palms, &c.; but care should be taken when planting, to allow sufficient space for each tree to develop, as mentioned in the preceding rules for gardens in general.

For a house standing far back in its own grounds, few things will look more inviting than a central avenue or drive, with a graceful bend, and planted with trees, and, if the shade is too great for grass to grow underneath, a line of tastefully-arranged rockery on each side, planted with maidenhair and other ferns, *Begonias, Caladiums, Callas, Tradescantia, Alocasias,* &c., cannot fail to have a good effect.

In gardens with a wide frontage, but with not space enough to admit of a central avenue, a half-circle drive might be formed, and the remaining space devoted to a lawn, tastefully laid out with flower-beds, palms, or specimen isolated plants, while a background could be made of trees and shrubs, and the borders devoted to annuals and dwarf flowering plants.

With larger gardens, a more ornamental system of undulating lawns, with clumps of shade trees, specimen palms, raised ornamental beds, serpentine paths, and rockeries, could be carried out.

In planting, care should be taken to have sunny spots and shady spots. This can be obtained by planting trees in clumps of threes and fives at angles to each other. Heavy permanent shade should be avoided, as it is conducive to draw weakly plants.

When planting, choose only straight healthy trees, and robust bushy shrubs, discarding all sickly specimens of which it is folly to expect will ever grow into nice plants.

Deciduous and evergreen trees require to be tastefully intermixed, so that the advantage of spring foliage and flowers may be obtained, and the continual sameness of a heavy green mass avoided. And again, in the formation and arrangement of shrubberies, care should be taken that the various forms and shades of foliage, both red and green, are tastefully intermixed, and that a blaze of red Acalyphas, or a continuous line of green shrubs, should not be planted together, but that just sufficient variety of plants are mixed to give a warmth of colour, and break the monotony of a continual line of green.

Large specimen palms in isolated positions, will in some cases be improved by having their bare stems covered with ivy, or other suitable climbing plants; and under the shade of large trees, where it is difficult to get grass to grow, a slight mound of rock-work and rich soil, filled with ferns and shade-loving plants, will impart a cheerful effect.

Paths.—It is generally the custom in Egyptian gardens, to pay little or no attention to the paths. One often finds them hollowed out, rather than raised, and after a heavy rainfall, they present miniature rivers, and are practically impassable. Even in the well-kept gardens of Cairo, this state of affairs can scarcely be said to be remedied, as the paths there are usually covered with small pebble quartz, which tends to give a somewhat finished appearance, but are exceedingly unpleasant to walk upon.

For private gardens, the main path should be from

three to four metres wide, and all paths that are frequently used should be wide enough to permit of two persons walking comfortably abreast, and to allow plants to be grown on either side. It is a great mistake to make paths too narrow, and this should be thought of when the garden is first laid out.

During the last three years the writer has endeavoured to introduce into the country, the raised paths which are so much admired in European gardens. Those in the garden of All Saints' Church, Ramleh, in the English cemetery at Alexandria, and in a few private gardens in the neighbourhood, have proved a great success, and have been universally admired.

For path-making, the soil, after being cut up and raked into shape, should be left just high enough in the centre to throw off the water, and to prevent it from wearing hollow, and the sides should slope gently off to the edge. The paths, when in shape, should be rolled with an iron roller, commencing on the outside, and continuing until a perfectly smooth and raised surface is obtained. Should the level of the path be not sufficiently high when the soil is cut up, broken stone, brick, or any rough material should be added until the required level is obtained, and the whole well watered and rolled. The foundation now being made, the path is ready for the surface materials, which can be formed by one of the two methods mentioned below, which will, of course, depend much upon the materials at hand.

(1st) Nile mud should be carted to the garden, and, if already partly firm, should be placed in heaps and treated in the same manner, as sand and lime is for making mortar, being banked up with a hollow in the centre. This hollow should be filled with water, and worked up by the Arabs with their feet and "fases" until the whole becomes a soft semi-liquid mud. The mud should then be carried in pails or oil tins, and spread over the paths either with a broom or by the hands, to the depth of about 2in.

Laying Out a Garden.

Next, broken pottery, clinkers, or any rough material, should be spread over and well rolled in while the mud is still soft, always commencing from the outside so as to keep the path in shape, and when the whole is dry, the work will present a compact, cement-like surface. Red clay or "tena" should then be poured on in a liquid state, and spread over thinly with a broom. Sand may be put on when the clay is partly dry, well rolled in, and any high places beaten down. For public gardens, &c., paths of this kind will be found to suit admirably, and will always present a tidy and neat appearance. Their construction is very rapid, three or four men when used to the work, will complete twenty to thirty metres per day, and with an occasional rolling the paths will last a lifetime.

(2nd) The other method is to shape the paths as previously described, and then to place the stones or clinkers on without the mud, covering them with soil which should be well watered and rolled. When this is set, the red clay, as already mentioned, should be poured over, and the sand rolled in before the clay is dry. This will prevent it from being blown away by the wind, and at the same time give a finished appearance.

This is a quicker method, but the former makes the stronger path.

Lawns.—Lawns to a garden are what carpets are to a house, and therefore the larger the area devoted to them, the better will be the appearance. Few things can present a cooler and more inviting effect after the heat and dust of an Eastern city, than an expanse of green lawn, while on the other hand a more unsightly object can scarcely be imagined than a conglomeration of fantastical-shaped beds, with ugly borders of red Alternanthera, and endless little paths, which are unfortunately so often met with here.

For lawn-making in Egypt four different plants can be used.

(1st) Lawn grass from seed.
(2nd) *Lippia nodiflora,* a low procumbent herb, known under the native name of " Libea."
(3rd) *Cyperus rotundus,* known as " Dis," a coarse grass-like plant with broad leaves, used for lawns in the Esbekieh Gardens at Cairo.
(4th) *Cynodon dactylon,* the native turf, called by the Arabs " Neguil," which is found growing on canal banks and damp places throughout the country.

Grass Lawns from Seed.—For undulated lawns, where a winter effect is to be considered, grass lawns from mixed seed is perhaps the best, as they present a cool bright green surface, which lasts from October to the beginning of June. It will then begin to dry in patches, and should consequently be cut up, and the soil allowed to remain fallow throughout the summer months.

About the beginning of September the lawns should receive a good dressing of well-decomposed horse-manure, and be levelled or undulated as required, care being taken to keep the surface a few inches above the path. The seed should be sown thickly about the end of the month, and be slightly covered with finely-sifted soil, and afterwards with a thin layer of well-rotted manure, and watered copiously with a fine rose on the hose. If the lawn is hosed three times a day, the surface will become green in from seven to nine days. As soon as the grass is tall enough, it should be cut with the reaping-hook or shears, and afterwards rolled to prevent it becoming spongy. Later on it ought to be cut every third or fourth day with the mowing machine, and rolled occasionally when dry.

Lippia nodiflora—*Libea.*—A low-growing, creeping herb of the Verbena family, found in damp places. It is extensively planted in the gardens of Cairo and Gizeh for lawn-making purposes. In Alexandria it is commonly used for summer or permanent lawns, though it is an undesirable plant in winter, when it becomes dull and patchy. In making lawns of this kind, the ground should be dug up and levelled in May or June, and small pieces of Libea planted about an inch apart until the space is covered. This should be frequently watered until the whole becomes a green surface. It requires going over occasionally with an iron roller, and care should be taken to cut off the flowers, which are of a dull-grey colour, with the mowing machine, or they will greatly tend to spoil the green effect.

Cyperus rotundus—*Dis.*—This plant can scarcely be recommended except for large areas, on account of its very coarse nature. It should be treated similarly to the Lippia, but attention should be paid to cutting and rolling, as the plants have a tendency to become knotty.

Cynodon dactylon.—*The Neguil*, or Indian doob-grass—a native grass known under the former name. It is of a trailing habit, vigorous in growth, and of a soft dark green hue. From experiments with this grass during the last three years in various positions, I have at length been able to introduce it for lawn-making purposes, into the Alexandrian and Ramleh gardens with great success, although a faint attempt, which ended only in a partial success, had previously been made.

The method found to be most successful is to first level the soil intended for the lawn, raising it an inch or so above the path, and well water it if dry. The Neguil should then be cut with the "fas" into small squares, of a convenient size, and laid face to face with the earth adhering to the roots. This prevents the grass from becoming dry, and makes it easier to pack in carts for conveyance to the garden.

Short, close-growing turf that has been eaten by cattle should always be chosen if possible, as poor thin turf rarely makes a good lawn.

On arriving at the garden the squares should be relaid closely together, and beaten with heavy wooden beaters until a uniform level surface is obtained. If the grass is a little long in places, it will be advisable to go over it with the mowing-machine at once, and any high places beaten down. Beating should never be done when the grass is wet, or great damage by bruising the turf will be the result.

After the first watering, a slight dressing of manure should from time to time be spread over the lawn. This will fill up any joins in the turves, and encourage the new grass to grow.

If lawns of this kind are made in February or March, and frequently watered, a beautiful dark green surface can be obtained in a fortnight.

For hot, sandy places near the sea, lawns of Neguil answer admirably, but they require to be cut and rolled, at least once a week during the summer months, and watered frequently, especially if they are required for tennis courts or other useful purposes, and the lawns should never be made under the shade of trees.

Owing to the increase on the demand during the last year for permanent grass lawns, some difficulty has been experienced in securing a sufficient quantity of suitable turf in the neighbourhood of Ramleh, and lawns are now being made from the seed of the same grass, which should be treated in a similar way to the lawn grass seed; but the dressing of manure should not be given until after the seed has germinated, and water must be put on lightly so as not to wash the seed into heaps. May is perhaps the best time for sowing, and with attention to weeding and watering a good lawn can be obtained in a month or six weeks.

Flower Beds.—The system of watering the garden by means of irrigation naturally necessitates low flower-beds and borders; but where the water can be given by the hose this need not be strictly adhered to. Slightly raised beds enable the plants to be seen to a much better advantage, and beds of this kind on undulated lawns always have a good effect.

Laying Out a Garden. 13

For spring bedding, many of the dwarf kinds of Roses, Pansies, Verbenas, Phlox, Carnations, Lilies, Freesias, Amaryllis, Arums, Mignonette, and many of the annuals can be tastefully arranged; while summer beds could be filled with Chrysanthemums, Petunias, Coleus, Iresine, Heliotrope, and many of the Zonal Pelargoniums make a good show. The lovely pink flowering Geranium (Souvenir de Chas. Turner), with a border of silvery-leaved *Centaurea candidissima*, has also a good effect. For late autumn, carpet-bedding, with the different varities of Alternanthera, *Gazania rigens*, Variegated Succulents, &c., would look pretty.

HEDGES.

Well kept hedges cannot fail to give a neat appearance to a garden, while badly kept ones always look neglected and untidy. This is, however, seldom the fault of the plants or the soil, but often depends in a great measure on the way in which the plants are cut or clipped. In the majority of cases, hedges are allowed to become overgrown at the top, and are invariably cut flat like a table. This prevents the sun and air from reaching the bottom, and thus the plants become ragged and full of gaps.

Hedges should always be cut so as to bring their tops up to a point; for example, a hedge 4ft. high should have 2ft. of its sides brought up straight like a wall, and the other 2ft. should be cut in a slanting direction on each side so as to form a pointed ridge. Hedges of this kind require to be cut as soon as they begin to grow out of shape, and, with attention, will always remain compact and tidy. The following are a few plants suitable for this purpose:—

Pittosporum undulatum.—This is one of the best plants for hedges or screens for a windy position. It has dark green leaves and clusters of sweet-scented white flowers in spring. For a further description see page 44.

Hibiscus rosa-sinensis. — A very suitable plant for hedges, with handsome foliage and large dark-red flowers. It strikes easily from cuttings in the autumn, and is much grown in the gardens of Cairo.

Duranta Plumieri—The Blembago of the Arabs. —A quick growing plant, with hanging clusters of pretty mauve flowers, thorny stems, and bunches of orange-coloured berries. It may be propagated either by seed or cuttings, and is one of the best plants for a hedge. It is an evergreen plant, but in very exposed positions it sometimes loses its leaves in winter.

Myrtus communis—The Common Myrtle.—This plant is often used for the purpose of hedges, its foliage and flowers having a pretty effect in spring, but unfortunately its leaves are often attacked by insect pests. It thrives best in rich alluvial soil.

Schinus terebinthifolia — The *Filfil.* — A useful plant for tall screens; it also lends itself readily to arching. To these may be added:

Ligustrum japonica, the "Privet"; *Viburnum odoratissimum; V. tinus; Lantana Camara*, and its varieties; *Nerium oleander; Justicia Betlonica; Laurus nobilis*, the "Sweet Bay"; *Elæagnus hortensis* (for screens); *Opuntia Ficus-Indica*, "Indian Fig"; and *Arundo Donax*, "Bamboo Grass," for windy positions. All the above-mentioned plants may be used either for hedges or screens.

CHAPTER III.

TREES.

THE large number of beautiful tropical and sub-tropical trees one finds growing in the country is now very considerable.

To the date palm, the essential feature of Egyptian scenery, are added many trees, some evergreen, while others are of a gorgeous flowering nature, which although representatives of the four quarters of the globe, yet they have for the most part fully established themselves in the country.

The value of trees in towns, for avenues and public resorts, is obvious in many ways, for not only do trees give shade, temper the heat, and render the surrounding atmosphere cool and moist, but they purify the air, by the carbonic acid gas they absorb, and the oxygen they give out—a fact that cannot be too much overrated in crowded Eastern cities; and it can hardly be from anything but ignorance of their value, that one so often sees large specimens in towns, growing in places where it would be difficult for a tree to again become established, cut and hacked about in the most barbarous manner, and in many cases removed altogether.

Much might also be done in utilising some of the more ornamental flowering trees for avenue purposes.

Planting.—This operation may be carried out with safety from the end of January, to the middle of

March, so that the trees may become well established before the rains have finished, though the work may be carried on still later if the season is favourable. If possible, transplant with a ball of earth at the roots, and press the soil firmly round them. In the case of the Poinciana, water and soil should be thrown in together. In all cases make the soil firm by having it well trodden about the roots, leaving a place to hold the water; and have the stems securely staked, to prevent them being broken by the wind. Large specimens on lawns, &c., may often be improved by having their branches cut back, so as to form a compact head; this will also lessen the danger of them being broken by the wind, and, in the case of large strong-growing trees, as Fici, &c., they will be greatly benefited by a trench being dug round their roots, about a metre in depth and width, and three or four metres from the trunk, at the same time cutting back all thick roots, after which the trench should be half filled with strong manure, night soil being preferred and mixed with the ordinary earth. The trench should be filled up with water for two days, and then level with the soil. This work, which can be carried out during the summer months, not only enables the tree to form fibrous feeding roots, but it prevents the larger ones from impoverishing parts of the garden with which they come in contact.

Acacia arabica, var. nilotica.—A native of Upper Egypt. It is a large, bushy, deciduous tree with delicate pinnate foliage which clothes the tree in April, and it blooms in June and July with numbers of small yellow flowers which smell like new mowed hay. The tree which is closely allied to the species producing "gum arabic," is known under the Arabic name of "Sunt." It is grown from seed in large quantities at The Barrage for planting on the sides of canals, but will not grow in Alexandria. Although the tree is admirably suited for binding the sand, it should never be planted so as to overhang, as the falling leaves and flowers, have a tendency to pollute the water, an example of which can be seen in June when the small canal which supplies Alexandria with water is turned yellow by the immense quantities of Lebbeck blossoms which fall into it from the overhanging trees.

Many of the exotic Acacias such as *A. Senegal, A. Adamsoni, A. dealbata, A. abyssinica, A. longifolia,* and *A. decipiens* are also grown in the more modern gardens of Alexandria and Cairo.

Albezzia Lebbeck.—A native of India and Ceylon. In Egypt it is the principal avenue tree. The foliage is deciduous, but the leaves remain on the tree for a long time if in a sheltered position. The new foliage appears in May, and the tassel-like flowers, known as "Dakn el Pasha," follow shortly after. The Lebbeck also flowers a second time in August. Its wood is much valued in the country for boat-building, water-wheels, &c., and large branches, or pieces of the trunk planted in winter or in the early spring, will the same year make nice trees. It can also be raised from seed, which germinates freely.

A. procera.—A tree similar to the above, though suitable for this climate, and growing freely from seed in Alexandria, is almost unknown. Its wood is even grained and durable, and is used for rice-pounders, sugar-cane-crushers, and agricultural implements. It would, if extensively planted, prove a valuable timber tree to the country.

Aliantus glandulosa.—A native of China, it has been introduced extensively into Algeria and Western Europe for feeding the silkworm. It is a rapid-growing deciduous tree, with large pinnate leaves, and the seeds hang in clusters after the leaves have fallen. The tree thrives in almost any soil, and propagates readily from seed.

Adansonia digitata—the Baobab Tree.—Native of Western Africa, where it grows to an immense size—one specimen is mentioned as having a stem

c

90ft. in circumference. It yields a bark suitable for paper-making, and small trees are sometimes met with in Cairo, where it is grown chiefly for the ornamental character of its flowers and foliage.

Balanites egyptica.—A large tree of a drooping habit, with small, dark foliage, commonly met with at inland places. It yields a quantity of small fruit of a brown colour, known under the name of Egyptian Myrabolans. The seeds, when fermented, are made into an intoxicating drink, called by the negroes "Zachum."

Bauhinia purpurea.—A pretty ornamental flowering tree of Cairo, with large kidney-shaped leaves, which close at night, and masses of purple pea-shaped flowers, which bloom on the whole length of the branches under the leaves, making it a beautiful object in April. The flowers are succeeded by long hanging pods, which remain until the following season. The tree is propagated by seed, but requires a sheltered position to develop, and flower. Other varieties, such as *B. variegata*, *B. reticulata*, *B. tomentosa*, and *B. aculeata*, are also grown.

Broussonetia papyrifera—The Paper Mulberry.—A common deciduous tree in Ramleh, with soft, downy leaves, and hairy fruits, nearly globular; about the size of a plum, scarlet when ripe, and of a sweet insipid taste. The bark furnishes the material known as "Tapa cloth."

Cordia Myxa.—The Monkheyt of the Arabs, who make a birdlime from the small cherry-like fruit. The tree makes a handsome specimen, with its slender hanging branches and large ovate leaves. It is naturalised in the country, and when once established, will grow under very arid conditions.

Cercis Siliquastrum. — Judas Tree, sometimes seen in the gardens of Alexandria and Cairo. It has curious, kidney-shaped leaves and beautiful pink flowers, which, unlike other plants, grow out of the bark of the stem and branches.

This tree shares with the Elder, the ignominy of being supposed to be the tree on which Iscariot hanged himself. The pods remain on the tree all the year, and are flat, thin, and brown in colour.

There is another variety, known under the name of *C. canadensis*, which has not yet been introduced into the country. Both are handsome, ornamental trees. Propagated by seeds.

Ceratonia Siliqua—The Carob, or St. John's Bread. —Native of the Levant. A branching tree, about 30ft. high, with shiny dark-green foliage, bearing pods which are known as locus beans. The dry saccharine pulp of these beans is very nutritious, and is supposed to have been the food of St. John in the wilderness. The small seeds are said to have served as the original carat weight of the jewellers. Propagated by seed and cuttings.

Cæsalpinia pulcherrima.—A native of the East Indies.—A showy tree with masses of small scarlet flowers and light fern-like foliage. The leaves are sometimes used as a substitute for senna.

C. Gilliesii and **C. Bonducilla** are also cultivated in Egyptian gardens as ornamental trees; propagated by seed.

Citharexylun cinereum—The West Indian Fiddle-wood.—Is a graceful tree of recent introduction, suitable either for the garden, or as an avenue shade tree. It grows rapidly from seed if planted in sheltered places, and bears in October thin hanging racemes of small white flowers.

Cassia fistula.—The handsomest of all the Cassias, it is a beautiful flowering tree, widely distributed throughout the tropics, where it grows from 30ft. to 50ft. high, having soft pinnate leaves, numerous racemes of bright yellow flowers, and cylindrical pendant pods, which often reach 2ft. in length. The pulp of the pod has a sweetish taste, and is used in medicine as a mild laxative.

Other varieties, such as *C. marylandica*, *C. bicapsularis*, and *C. corymbosa*, might here be mentioned as sometimes being seen in the gardens. All are propagated from seed.

CONIFERÆ.

Under the heading of this large natural order of the pine family, which is divided into six tribes, valuable representatives of each, remarkable for their graceful habit, and evergreen foliage, are fully established in the country, many of which form handsome trees. All can be raised from seed.

TRIBE I.—CUPREOSINEÆ.

Cupressus sempervirens—Cypress wood.—So-called from the aromatic and almost imperishable nature of its wood. It is a tall compact tree, of pyramidical form and dark green foliage; a native of Italy, and easily grown from seed. It is commonly met with in Alexandria, and will thrive in very poor soil.

C. fastigiata. — A tall, heavy-looking tree, often planted for avenues in Italian gardens.

C. horizontalis.—An allied species; is of a more spreading habit, like the cedar; is also used for avenues in this country.

Callistris quadrivalvis — The Sandarach. — The powdered resin of which furnishes "pounce," used in the preparation of parchment; is also one of the trees comprising Tribe I. found in the Egyptian gardens.

TRIBE II.—TAXODICÆ.

Taxodium distichum — The Deciduous Cypress.— This is a tall, graceful tree, with light, spreading foliage, small specimens of which are sometimes met with in the Ramleh Gardens. It is closely allied to the "Wellingtonias," the mammoth trees of California. *Cryptomaria japonica*, with which we are familiar in our gardens at home, has also been lately introduced into Egypt.

TRIBE III.—TAXEÆ.

Salisburia adiantifolia—The Maidenhair Tree.— This handsome tree, with its large fern-like leaf (as the name implies), has flowers resembling those of the common Berberry and a small acid fruit, it is, unfortunately, rarely met with, though I am informed that a very fine specimen was cut down a few years ago in the Government gardens on the Mahmoudieh Canal. *Taxus baccata*, the English Yew, and *T. fastigiata*, the Irish Yew, are seldom seen—except as small specimens in pots.

TRIBE IV.—PODOCARPEÆ.

Podocarpus Totara.—A valuable, hard-wooded evergreen tree of New Zealand, with small box-like leaves, and of remarkably slow growth. A few large specimens exist in the gardens near the canal at Alexandria, and small plants in pots are offered for sale by the local nurserymen.

TRIBE V.—ARAUCARIEÆ.

A genus of exceedingly handsome trees, very ornamental when growing upon a lawn, where there is ample space for the lower branches to develop. They can be raised in sandy soil, from cuttings taken from the half-ripe terminal side shoots; but some species will be found to strike more readily than others. The commonest method adopted in Egypt for raising young plants is, however, by means of seed, which must have previously been impregnated by the male plant. The seeds, which are contained in cones, should be sown soon after ripening, as otherwise they quickly lose their vitality. Sow each seed singly in small pots filled with well-washed sand, and place in shelter undei glass until it germinates. After the plants have made a little growth, they should be plunged, with their pots into the soil, in a half shady position, and potted on until they become large enough to plant out permanently in the open ground.

Araucaria excelsa.—The Norfolk Island Pine.

This variety is perhaps the best of the Araucarias, and is said by some authorities to be unrivalled in its beauty by any other tree. Tall specimens are sometimes seen in the gardens inland, but only small plants are met with in Alexandria.

A. Cookii.—A few trees of this variety still exist in the Esbekceyeh Gardens at Cairo. It makes a fine specimen, but has thicker branches, and it is not so dense in its growth as the former. It is a native of New Caledonia.

A. Bidwillii.—The Moreton Bay Pine. This handsome tree is quite distinct from the two preceding, having prickly, sparkling, dark green leaves, with spreading branches, rising from the ground in a pyramidical form. Its foliage is in character similar to the *A. imbricata* (the monkey puzzle) so often seen in English Gardens, but this latter variety is unknown in Egypt.

A. Cunninghamii.—A native of Australia. A vigorous growing species, conspicuous in Ramleh and Alexandria, where it towers above all other trees. It has a dark thick foliage of a cypress-like character, and has the advantage of thriving on dry, sandy soil in exposed windy positions where scarcely any other tree will exist. It requires very little water when once established, and always grows in an upright and well-balanced form.

This tree can be raised in sandy soil from seed or from cuttings taken from the roots 3in. or 4in. long. If the latter method is adopted, care should be taken that the cuttings are planted deeply in sandy soil.

TRIBE VI.—ABIETINEÆ—PINES.

Pinus halepensis—the Aleppo pine, *Senoubar*—a native of South Europe.—This species is fairly common in Egyptian gardens, where it makes an excellent avenue tree, and imparts a good effect when planted in clumps.

P. longifolia.—A resin-yielding pine of the Himalayas. It is a graceful species, with long needle-like leaves. Large trees are not often met with, but seeds received from Saharanpur, India, grew freely in the open ground at Alexandria.

P. pinaster—The Maritime pine—and **P. pinea**—The Stone pine—which has edible seeds, are common in the country.

P. elata.—A dwarf species, only rising a few feet high; is sometimes used in Alexandria for the centre of large flower-beds.

Euphorbia pulcherrima—Poinsettia—*Bint-el-soul.*—A familiar object in Egyptian gardens. Its crimson bracteal leaves, forming large star-like heads, makes it conspicuous at Christmas time. The plant strikes freely from cuttings, and blooms on the new wood. It should be cut back to a bud or two from the stem after flowering, and should be planted in a sheltered position, so that its beauty is not destroyed by the wind. Specimens trained in standard form have also a good effect.

E. pulcherrima, var. albida.—A white variety is sometimes grown, but is of comparatively little beauty.

E. neriifolia and **E. tirucalli** are also suitable for a shrubbery and grow into fine specimens.

E. jacquiniflora is a shrub rather than a tree, bearing a profusion of vermilion flowers, and slender thorny stems. It is a pretty plant for a dry sunny spot, or as a pot plant.

Erythrina crista-galli—The Lobster Plant.—A showy deciduous tree, with ternate leaves and clusters of claw-like flowers, of a deep crimson colour growing on the young wood. These flowers continue in bloom throughout the spring and summer. All the previous year's growth requires pruning back to the old wood, which should be done in January.

E. Corallodendron—The Red Bean Tree.—A small bushy tree with large sprays of brilliant red flowers, which open before the leaves appear. It grows freely in Cairo, where it a striking object in February and March.

E. Indica.—A larger specimen than the two preceding trees, common throughout Egypt, bearing in Spring, before the leaves appear, clusters of scarlet flowers, at the end of its branches. It is one of the trees that are grown to give shade to vanilla plantations. Most of the Erythrinas strike freely from cuttings in the early Spring.

Eucalyptus globulus.—A large tree in Cairo, often bearing two distinct kinds of leaves. The *Eucalypti*, unlike all other hardwooded trees, are remarkable for the rapidity of their growth, and the large quantity of water they absorb; hence they are admirably suited for planting in damp, malarious, districts.

Amongst them rank the tallest trees in the world.

E. resinifolia, E. amygdalina, and many others grow freely from seed, and thrive in sheltered places.

Ficus.—For the purpose of shade, few trees are more suitable than the many species of Fici, and as their wants and requirements are comparatively small, they readily find a place in almost every Egyptian garden. Some varieties propagate easily from cuttings, or from seed, but the best and quickest method is by taking the suitable branches and making a cut at a node or joint, half-way through the wood, in an upward direction. The cut should be kept from healing, by having a small piece of tile or stone placed within it, and a divided flower-pot or tin vessel filled with moist sandy soil, should be securely tied round the portion of the branch operated upon, and kept moist.

This is one of the safest ways of propagation, and may be applied to all trees and shrubs that are difficult to strike from cuttings. It invariably proves successful, and the branch being only partly cut through, allows the sap to ascend, and thus the upper part is kept alive, whilst the cut, arresting the sap as it descends, causes fresh roots to form in the moist soil in the pot, and in this way large plants may be obtained in a few months. Spring is perhaps the best time for the operation, although it can be carried on all through the summer.

F. macrophylla.—A native of Australia. One of the handsomest trees of the country, with large, oblong leaves, pointed at the apex, and terminal white sheaths. This species is similar in form to *F. elastica* (The Assam rubber), but it makes a more compact tree than the latter, which is distinguished from the *macrophylla* by its terminal red sheath which enfolds the young leaves.

F. nymphæifolia—The water-lily leaf Ficus.—A South American variety, of a bold, spreading habit, and large, ovate leaves. Some very fine specimens may be seen growing in the Esbekeeyeh Gardens at Cairo, but in Alexandria it seldom grows into a large tree.

F. eriobotryoides—The Loquat-leaf Ficus.—Commonly seen near the coast. It is of spreading habit, with tuffs of oblong leaves, having a rough brown under-surface, and hairy stems. It will grow on very poor soil, and would serve as an excellent shade tree for a windy garden. It will, however, lose its leaves if planted in a very exposed position, by the strong spring gales.

F. bengalensis—The Banyan tree.—A well-known Indian species, easily recognised by its curious aerial roots, which hang from the branches in large numbers.

The effect of these trees in a large garden is exceedingly tropical, as they form in themselves a complete continuity of shade, and the area they cover in the tropics by means of their hanging roots, which support the branches, is very considerable. Many fine specimens exist in Egypt, and it is recommended by some authorities that all superfluous roots should be removed.

F. rubiginosa.—This variety has smaller leaves than the Banyan, but it also has a tendency to send out bunches of aerial roots from the upper branches, though very much smaller than the former. It grows into a dense compact tree.

F. Benjamina.—A graceful drooping variety, with small glossy leaves, which gives it an excellent appearance as a specimen for a lawn. It grows well in Alexandria, but is not often planted.

F. retusa, var. nitida.—Also a handsome small-leaved variety. It is commoner than the former, and can either be trained on lattice work as a climber or grown in the form of a tree.

F. religiosa.— A sacred tree in India. It is a diciduous species, having hanging cordate leaves with a long tapering apex like a poplar.

F. Sycomorus—The Sycamore Tree—*Gaimmayze*. —This tree produces its fruit on the trunk and thick branches. It often serves for avenue purposes, on account of its spreading habit and the shade it gives. The fruit is eaten by the Arabs, and the coffins of the ancient Egyptians are said to have been made from its wood. The tree is a native of Egypt and is identical with the Sycamore of Scripture.

F. infectoria.—A West Indian variety, is used in Cairo as an avenue tree. It is remarkable for its leaves, which, turning yellow, fall, and new ones appear in the space of a week or so.

F. japonica.—The real name of which is yet undetermined; is also used as an avenue tree. It has leaves like the Mango, and is usually pruned annually into a bushy head. Many other species of a more or less ornamental value, such as *F. Gibbosa, F. parisitica, P. Parcelii, &c.*, are also frequently met with, but space will only admit of mention being made of the chief.

Jacaranda mimosifoli.—A native of tropical South America. This tree, with its masses of blue flowers, ranks with many only second in beauty to the red *Poinciana*. Its foliage is similar, only a little darker in colour, but the habit is more erect. It blooms a month earlier, that is in May, and often before the leaves are fully expanded. Propagated by seed, but is difficult to transplant when established, and will only flower in a sheltered position. For this purpose in windy gardens a large evergreen tree should be grown on the exposed side.

Kigelia pinnata.—The Sausage Tree of Nubia.— A tall tree in Cairo, peculiar on account of its long hanging fruits. It has coarse pinnate leaves, and flowers which only open at night.

Lagunaria Patersonii—The Norfolk Island Oak.— A tall, pyramidical-shaped tree, with small leaves, white on the under surface, and thick petaled pink flowers like an *Hibiscus*, to which family it belongs. A suitable tree for poor soil and a windy position.

Melia Azedarach—The Persian Lilac—*Zenzillia.*— An ornamental tree, often seen in Egypt, with pinnate, deciduous foliage, and clusters of small, lilac-coloured flowers. The tree grows rapidly from seed, but has a tendency to become straggling if not pruned. The fruit yields an oil; the bark contains medicinal properties; the leaves are used for poultices; and the trunk is valuable as timber.

Poinciana regia.—A native of Madagascar. One of the most gorgeous of all Egyptian flowering trees. It is striking both in its foliage and in its flowers, which makes it a general favourite. The graceful, fern-like leaves appear in May, and the whole tree is a mass of bright-red bloom a month later; a second flush of bloom will also appear (as with many other plants) at the rise of the Nile, in August.

The tree is raised from seed, which usually flowers in about the sixth or eighth year, and then only in a sheltered position. Its spreading, umbrella-shaped form requires but little pruning.

Plumeria acutifolia—The Indian Jasmine.—A dwarf spreading, soft-wooded tree, of the same family as the Oleander *(Apocynaceæ)*. It has deciduous, lanceolated leaves, and terminal bunches of sweet-smelling flowers. There are two varieties : *P. alba*, which has white flowers, with creamy-yellow centres ; and *P. rosea*, which has flowers of a deep pink. The plant is suitable for a dry position, and strikes easily from cuttings.

Parkinsonia aculeata.—A common tree in Cairo. Its light green foliage and showy yellow flowers makes it a pretty tree in April.

Salix babylonica—The Weeping Willow—*Safsaf*.—A very graceful tree for the borders of ornamental waters or lakes. Its drooping form and light green foliage makes it a striking contrast to the surrounding trees. It is common in Cairo, but is not often met with in the gardens on the coast, possibly on account of the sandy soil, which does not suit it, and through some unknown cause it is apt to die off suddenly.

S. Ægyptiaca is largely grown for binding the banks of canals.

Schinus molle.—The Drooping Filfil—Is commonly met with throughout the country. It is well suited for the same purpose as *Salix Babylonica*.

Tecoma stans.—A familiar ornamental tree, with drooping clusters of yellow, funnel-shaped flowers, which continues to bloom almost throughout the whole year in sheltered spots.

T. Cavendishii.—A dwarf tree with thin sessile leaves, and single tubular flowers. It has lately been introduced from India into the Ramleh Gardens.

Many other beautiful trees, as *Casuarina equisetifolia*, *Magnolia grandiflora*, *Phytolacca dioeca*, *Wigandia caracasana*, &c., &c., are of course crowded out for want of space.

CHAPTER IV.

PALMS—NAKHLA.

THE value of palms can hardly be over estimated, their varied forms, the graceful character of their foliage, and the tropical effect they impart, either in groups or as isolated specimens, cannot fail to make them popular.

With several of the commoner varieties a considerable amount of their beauty is lost when they attain a certain height, by the lower portion of their foliage alone being seen; still, to many who are familiar with them only as small specimens in pots, a tall stately palm will always prove an object of admiration, and their cool and pleasing appearance when in health, together with their easy culture, will amply repay the little care that is necessary to bestow upon them.

The majority of the ornamental varieties bear seed freely, and are therefore easily propagated by sowing the seed in the spring or the autumn. Some of the species take a considerable time to germinate, and strict attention should be paid to draining the pots or pans, as sourness of the soil is fatal to the young seedlings. With large specimens in the open ground too much water will often cause the foliage to turn yellow, especially in sandy soil, while dryness and cold winds are equally ruinous to the foliage.

Palms, unlike many other plants, if kept without water do not show the ill-effects at once, but in a few days the leaves will commence to turn yellow. This can never be rectified, the injured leaves will continue to lose their green colour, and must be cut off, and the plant stood in a half-shady position until the new foliage appears.

Constant attention should be paid to the watering of palms in pots for the verandah or the house, and if possible a few plants should always be kept in reserve in order to permit a weekly changing, and thus preventing those in use from becoming drawn and sickly. For this purpose a light wooden framework, slightly shaded with date palm leaves, will be found very useful in protecting the plants, whose leaves have become tender by standing in the house, from being scorched by the sun.

Specimen palms on lawns will be greatly benefited during the summer months by occasional supplies of weak manure and soot-water, which should never be given to any plant, particularly those in pots, when the soil is very dry, for if this is done burning of the roots will be the consequence.

The following list will be found to contain many of the species worthy of cultivation in the gardens here :—

Phœnix dactylifera—Date Palm—*Nakle*. — The Date Palm is one of the few trees indigenous to the country, though the best quality of fruit is said to be the "Lady's-finger date" of Tunis. The long black date, the typical type of the genus known as *Hayameh*, is one of the most popular.

Professor Sickenburger, a well-known authority, once said that there are twenty-seven varieties of Date Palms, different in size, shape, and colour, and each peculiar to the locality in which they grow. These are divided into two classes—those suitable for preserves, and those which from their fermenting properties must be eaten at once. The palms which are of either sex (diœcious) are usually propagated by offshoots which grow at the foot of the parent palm.

These are disconnected after they have formed roots, which is usually about the fourth year, and with proper attention to watering and manuring they will fruit in the eighth or tenth year.

Planting.—The best time for transplanting is about the first or second week in February, and large palms 20ft. to 30ft. high can then be shifted with safety. When transplanting, these palms should have their stems placed from two-and-a-half to four metres in the soil, the depth being in proportion to the height of the palm. This is done in order that they may root up the stem, and the water should be allowed to run in with the soil, so that the whole may become a solid and compact mass. Tall specimens will, in many cases, require supporting for a time, in order that they may settle completely upright.

In preparing the palms for shifting, two or three rows of the outer leaves must be cut off, and the remainder fastened together, and tied round with a mat. This is done to protect the heart, which is the vital part of the palm, from becoming dry by the sun. Abundance of water should be given once a week for the first three months, and fortnightly for the three months following. About the end of May the palms should be visited and any flower spikes cut off. Loosen the mat slightly as the leaves expand, otherwise the heart is liable to become weak, and when the mat is removed it is easily snapped off by the wind and the palm destroyed. As the leaves strengthen, the loosening should be gradually continued until the mat is removed altogether about the middle of September. Palms growing on unirrigated ground are more likely to succeed when transplanted, than those growing on irrigated soil.

Date Palms flower in April, and require to be artificially fertilised, which is done by taking a small bunch of the staminate flowers, and placing them amongst the pistilate. They should then be tied to prevent the wind from blowing them away. For this purpose it is usual to employ a native who understands the work. As the fruit ripens the bunches should be looked to, any decayed or deformed fruit picked out, and in some cases netted to protect the dates from birds and bats.

ORNAMENTAL PALMS.

Areca Baueri. — A handsome species requiring shade. It has large pinnate arched leaves, and is a suitable palm for a large conservatory or the varandah. Pot in sandy loam and manure.

A. lutescens.—An excellent palm for a clump on a lawn or in a moist place, throwing up several cane-like stems which are yellow and spotted, and light pinnate leaves. It also makes a useful pot plant, but should not be exposed to the wind in winter.

Astrocaryum mexicana.—A fine specimen palm, allied to the Cocos *(Cocoinæ)*, with coarse pinnate leaves, white on the under surface, and flat black spines which will cause the flesh to fester if come in contact with. It can be propagated by seed, and large plants in tubs seed readily, but being of an exotic nature, it requires the shelter of a glass conservatory in winter.

Caryota urens—The Wine, or Toddy Palm; native of India and Ceylon.—The Toddy, an intoxicating drink, is made from the sap, and a good tree is said to yield a hundred pints in twenty-four hours. In Alexandria small plants are sometimes seen in sheltered places, but in Cairo it assumes a tall, tree-like form with a brown slender stem, and a crown of spreading leaves like a large maidenhair fern. Propagated by seed and suckers.

Cocos flexuosa. — This variety is one of the handsomest palms in the country, clumps of which often form a prominent feature in the gardens of Cairo. Smaller plants are also occasionally met with in Alexandria. It has a tall, smooth stem, swollen slightly at the base, and a massive head of feather-like leaves. It is one of the best palms for an avenue, and if planted in groups of five or seven, it has a rich tropical effect in a large garden.

Cocos nucifera—The Cocoa-nut Palm—*Groz-zel-Ind.*—This species is, without exception, the most extensively cultivated tree in the tropics. When cultivated, it begins to form a stem and bear fruit about the eighth year, and continuing from seventy to eighty years, growing from 40ft. to 80ft. in height, and bearing annually from thirty to sixty nuts. It grows best near the sea in salty soil, but the winter temperature of Alexandria is too severe for it. A specimen plant in a sheltered part of the garden of the Alexandria Water Works, has stood outside for several years. For pot culture it makes an excellent specimen plant, and if protected in winter it will last for several years. For this purpose, the nuts should be obtained from the bazaars, those being chosen which have their kernel shooting out. These should be planted in pots with the shoot just level with the soil, stood in a shady position, and watered sparingly until the leaves appear. The palms will form nice specimens by the second year, the leaves being partly fan-shaped, with yellow stems, but becoming pinnate as the plant increases in age. A little salt-water may be given occasionally.

C. plumosa.—A very ornamental species with light drooping foliage. Small plants are very useful as a background in glasshouses.

C. Weddeliana.—A dwarf variety, with a slender stem and graceful arched leaves. It is perhaps the handsomest pot palm for a greenhouse, but is somewhat difficult to cultivate.

Chamædorea, sp.—A climbing palm with a slender notched stem and a crown of broadly pinnate leaves at the top. Small plants in pots are only met with.

Chamærops Fortunei— The European Fan Palm.— This is, perhaps, the hardiest species of the Palmeæ,

D

growing, as it does, all the year round in the grounds of the Royal Gardens at Kew. It is a pretty dwarf palm, easily cultivated, with spikey stems, and deeply-cut leaves. This species often has a number of offsets growing at the foot of the plant.

Hyophorbe amaricaulis.—A very ornamental variety, much resembling an *Areca*, with a cylindrical reddish-brown stem, and slightly-arched pinnate leaves. A suitable palm either as a specimen for the open ground or for pot culture. Propagated by seed.

Hyphæne Thebaica — The Doum Palm. — This species, which is characteristic of Upper Egypt, is remarkable for having normally a branched trunk, instead of a single one with a terminal crown of leaves. It has a wide range in eastern tropical and sub-tropical Africa, and occurs also in Arabia. Its brown, fibrous fruits, which have the taste of ginger-bread, are eaten by the natives, though to the European taste they are anything but palatable. A small plant of this species may be seen in Cairo, but it is not met with further north.

Kentia Fosteriana—The Thatch-leaf Palm.—This plant is suitable either for pot culture, for the plant-house, verandah, or as a garden palm, growing as it does in the open ground all the year round in Alexandria. It is of a spreading habit, with glossy leaves and slender stems. It is propagated from seeds.

Other varieties, as *K. Belmoreana*, *K. Canterburyana*, and *K. Australis*, are also grown.

Latania borbonica, syn. *Livistona chinensis*—the common Fan Palm.—This species is well known in the country, being largely grown, both as a pot-plant and also as a specimen palm, in the open ground. It is a very hardy species, growing in any kind of soil, but the leaves should be tied up during the winter in very exposed places.

L. aurea and **L. rubra** are cultivated in pots, for the plant house.

Metroxylon sagus—The Sago Palm.—This species is occasionally seen in the gardens of Cairo. It has pinnate leaves and erect petioles, but its somewhat slow growth prevents it from being a popular garden palm.

Phœnix canariensis.—Of the many ornamental varieties of this genus, the above-named species is, perhaps, the best in the gardens here. Its handsome head, formed of a thick growth of pinnate, arched leaves, makes it a very desirable specimen plant. Some very good specimens grow in the ordinary garden soil of Alexandria.

P. leonensis.—A thick-stemmed variety, having an immense head of small pinnate leaves with thick petioles, which radiate out in a reflex form like a huge *Cycas*, making it a striking and excellent shady specimen. A very fine palm of this variety can be seen in the garden of Mr. Cicolani, at Cairo. The plant is somewhat rare.

P. rupicola.—An Indian variety; is also a desirable acquisition on account of its elegant growth and graceful drooping leaves. Other varieties, suitable either for pot culture or the open ground, are *P. acaulis*, *P. reclinata*, *P. spinosa*, *P. sylvestris*, *P. zeylanica*.

Rhapis flabelliformis—The Ratten Palm.—A dwarf fan-shaped palm, native of Eastern Asia. The leaves, which are deeply cut into segments, grow on long thin stems, forming a dense clump and imparting a tropical effect, which renders the palm valuable for planting, either singly, or in groups of three, on large lawns. The plant is fairly hardy, and will stand a certain amount of wind. It can be propagated by offsets.

Sabal Blackburniana—The Bermuda Fan-Palm.— A large, handsome species, growing to the height of 20ft. to 30ft., and having clusters of round and black seeds. The leaves are characteristic, as showing the intermediate stage between the fan-leaved and the pinnatifid-leaved palm. It is often met with in the small public gardens of Cairo, but large specimens are rare.

S. princeps.—A smaller species, with long, upright leaf-stalks. It is, perhaps, commoner than the above-named variety, but not so bold in character. Some nice plants may be seen in the gardens of Shepherd's Hotel, at Cairo.

Seaforthia elegans.—A native of New South Wales; grows well in the gardens on the Mahmoudieh Canal. It has a smooth, ringed stem and heavy drooping pinnate leaves, which often hang in an irregular untidy form. It is, perhaps, seen at its best as a tall specimen.

Thrinax parviflora.—A very handsome pot palm for the greenhouse, having fan-shaped leaves divided into a star-like form.

T. elegans, *T. argentea*, *T. elegantissima*, *T. gracillima*, and *T. glauca* are suitable palms for the conservatory.

Washingtonia filifera—The Californian Fan-Palm.—A very handsome, hardy species. It is distinguished from the *Latania* by the long thread-like filiments which hang from its large fan-shaped leaves. It forms a noble plant in very poor soil, and if the leaves are tied up in the winter to prevent them being broken by the wind, it will be found to be a suitable palm for gardens on the coast. It can be raised from seeds, or young plants purchased from the local nurserymen.

PANDANEÆ.

Pandanus odoratissimus—The Screw Pine.—A very handsome plant, suitable for a large garden. It has terminal heads of long sedge-like leaves arranged in a screw-like manner, from which it derives its name, and the large aërial roots which hang down from the branches gives it a curious effect. The plant, which is of two sex, flowers in March and April with huge panicles of creamy-white flowers, which impart a strong fragrance like pine-apple. The fruit is in the form of a large cone. It is commonly met with in the gardens of Alexandria and Cairo, and can be propagated by cuttings.

P. Veitchii.—A very decorative dwarf variety, with striped leaves. It is usually grown in pots, as it requires protection during the winter, when care should be taken that the water does not lodge in the leaves, which will cause the heart to rot and destroy the plant. It is easily propagated by suckers, which are borne in great abundance on the old plants.

Other varieties cultivated for decorative purposes are: *P. bifurcatus, P. graminifolius, P. ornatus, P. spectabilis,* and *P. utilis.*

Cycads.—A curious genus of plants, resembling a fern and a palm, but bearing a cone-like fruit. They are representatives of a very ancient flora. The leaves which are long and pinnate, radiate from the stem and form a complete circle. The crown is exceedingly handsome when the young leaves are appearing in the spring, at which time a little manure water will be of help to their roots. Some varieties are propagated from suckers and seeds, but being of slow growth it is usual to import the stems. The plants are very subject to the attacks of scale and other insect pests, and require constantly looking over. Pot in sandy loam, broken brick, and leaf soil.

Cycas revoluta.—A native of Japan, has now become naturalised here. It is a very hardy plant, and can be propagated from suckers.

C. circinalis.—This plant is of a more delicate growth, with longer leaves of a polished light green. Plants that have become sickly should be taken out of the soil, and any decayed parts at the roots and base of the stem, scraped off. The stems, after being dried in the sun for a day or two, should be then repotted and placed or plunged in the soil.

Other varieties as *Encephalartos Caffra, E. horridus, Zamia integrifolia, Z. Lindenii, Macrozamia mackenzii, M. plumosa,* and *M. corallipes* might with advantage be introduced for cultivation.

CHAPTER V.

SHRUBS.

To the formation and arrangement of shrubberies I will refer my readers to the remarks on "Laying out a Garden," page 5.

The transplanting season for all plants, as previously mentioned, is from the middle of January to the middle or end of March; but the first or second week in February is the best time for the removal of choice specimens.

Previous to the shifting of the shrub, the ground should be as dry as possible, in order to prevent the ball of earth round the roots from breaking, and a trench 2ft. wide should be dug round the plant, leaving the ball of earth and roots in the centre. Any roots that project out should be cut off with a sharp knife, and when a sufficient depth has been attained, the ball should be carefully undermined with a trowel. When the soil has been dug out half-way, a piece of strong canvas or matting, about a metre and a half square, should be half rolled up and placed under the ball, and with a little pressure the plant will heave over, and the canvas can then be drawn through, tied round, and the plant carefully lifted out and carried to the place required, where the hole should have previously been dug to receive it.

When planting, see that the shrub is perfectly upright, and have the neck of the plant level with the surface; untie the canvas, and let it fall round the bottom of the roots, as the ball may be broken, and the plant lost if attempts are made to remove it. When filling in the soil press it firmly, making a trough round the top, which should be filled daily with water. Any branches that have been tied up may now be undone, and the foliage occasionally syringed. Large plants may be shifted in this manner without any ill effects.

In removing plants in sandy soil, where it is difficult to shift them with balls of earth at their roots, it will be found to be of great benefit to the plants, after transplanting, if a hose with a fine rose attached, is kept playing for a week or so over the foliage during the day until new roots are made, and the borders should he mulched with manure, to prevent evaporation. The manure can afterwards be dug in.

New borders for shrubs should be dug deeply, and if the soil is poor and sandy, Nile soil and manure should be added. Where the ground is of a heavy nature, as one often finds in Cairo, a good dressing of sand or ashes will help to keep the soil open.

Plant always in angles, and consider the habit and growth of each plant, having the taller behind, and those of a dwarfer nature in front, allowing sufficient room for each plant to develop without touching its neighbour.

Where the shrubbery does not adjoin a lawn, a verge or strip of turf about 3ft. wide, or a border of suitable dwarf plants, such as *Gazania rigens*, which blooms with large yellow flowers from April until July; *Cineraria maritima*, *Tradescantia tricolor*, and others would give a tidy appearance, while sufficient room between the shrubs and the adjoining plants should be left for Dahlias, Lilies, Sweet Peas, Carnations, Poppies, &c., so that the borders may present banks of foliage and flowers. All shrubs should be pruned into shape at least once a year, usually after flowering.

The following are some of the best varieties of shrubs suitable for the usual requirements :—

Aralia digitata.— A large bushy shrub, known under this name, found in almost every garden in the country. It is a nice plant when well grown, with its glossy digitate leaves, and it bears during the summer hanging masses of brown and yellow flowers. It propagates easily from cuttings, and is largely grown as a pot plant for decorative purposes.

A. Veitchii.—A handsome slender species, with finely-cut leaves of a brownish colour. It makes a nice plant for the conservatory, where it should be grown.

A. filicifolia.—A greenhouse variety with green serrate leaves, usually grown for stocks for grafting the *Veitchii*. It can be propagated by cuttings.

A. pentaphylla.—This variety is the best in the country, having large, horse-chestnut-like leaves, which curve in a reflex form, and grows in sheltered positions into a handsome, bushy tree.

A. papyrifera—The Chinese Rice-Paper Plant.—This species is common in the gardens of Alexandria and Ramleh, where it usually grows on a single stem, with a cluster of white, hairy leaves at the top. It can be propagated by seed or suckers; the latter are thrown up in large quantities round the plant.

Acalypha Wilkesiana.—A large red-leaved shrub, common in the gardens here. It grows into a handsome compact bush when cut into a pyramidical form, and is suitable either for the background of a border, or as a specimen plant for a lawn. It thrives also in a shady position, but requires the sun to bring out its colour.

A. hispida. — A small-leaved variety with dark green centres and red margins. A nice plant for a small hedge.

A. marginata. — A similar plant to the above variety, but with larger leaves. All Acalyphas are easily propagated from cuttings, and in a sheltered position retain their foliage throughout the year.

Bambusa arundinacea—The Bamboo.—The tall, plumy stems of this plant have a very handsome effect in a garden. They are either suitable for a screen or for growing in clumps; but they require considerable room to spread, a good rich soil, and plenty of water.

For multiplying, large pieces of the roots, which contain eyes, should be sawn, or chopped off, in March or April, and planted in well-manured soil. Give a mulching of manure after planting, and water should be applied twice weekly. The first year the growth will be rather thin, but by the second year the plants will become established, and throw up stout canes, which grow very rapidly.

Care must be taken that the canes in bending over with the wind do not touch a wall, or great damage will be the result. Several dwarf varieties, as *B. japonica*, *B. nana*, and *B. nigra*, are grown for shrubberies, but only one species, namely *B. abyssinica*, is known to be indigenous in the whole of Africa.

Biota orientalis—*Afs.* — A dwarf Cupressus-like evergreen, growing in any kind of soil, and forming a compact, handsome shrub. It can be raised from seed, and nice plants obtained in three years, but some will be found to be of a less compact habit than others. They can be utilised either as dwarf hedges, isolated specimens, or pot plants.

Croton.—An exceedingly handsome genus of ornamental foliage plants, popular with all on account of the rich and beautiful markings of their leaves. A large number of named varieties are at present in cultivation in Europe, and new ones are constantly appearing; but only a few of the stronger kinds succeed in the open ground in Egypt, and then only in sheltered gardens, while the majority require the protection of a glasshouse or a warm sheltered spot in the winter.

C. aucubæfolius-superbus.—This is one of the best and hardiest varieties, growing into a large handsome bush in the open ground in a half-shady position. It has small green leaves covered with bright yellow spots, and is an excellent plant either for verandah or house decoration.

C. aucubæfolius-giganteus.—This is also a hardy variety, similar in habit and growth to the above-mentioned species, but with larger leaves.

C. Veitchii.—A well-known variety, with large oblong leaves and red veins. It is usually met with as a pot plant, and often grows into a fine specimen. Several other varieties stand out during the whole year in Cairo, where the winter temperature is higher than in Alexandria; but with the exception of the three here mentioned, all Crotons should be protected during the winter in gardens on the coast.

Culture.—For the successful culture of a large collection of Crotons, a sandy rich soil, composed chiefly of well decomposed vegetable matter, or potting soil (see page 70) well mixed with sand is a suitable compost. The pots should be clean, and well-drained by placing a large convex piece of pottery over the hole inside the pot, and 2in. of smaller pieces on top. See that the hole is sufficiently large enough for the water to pass out freely, or the soil will quickly become sour. Pot firmly by pressing the soil round the roots, and leave an inch or so at the top of the pot to hold the water.

March is a good month for potting, but the plants should not be shifted unless the pot is full of roots or the soil is sour, and then only shift into a vessel a size larger. Healthy growing specimens will require to be shifted again about the end of August. After the spring potting, they should be placed in the glasshouse or a shady position, and the foliage syringed occasionally until the plant has made new roots. Air should be given on all possible occasions, or loss of foliage, *thrip*, and other insect pests will be the consequence.

About the middle or end of June, the plants should be plunged with their pots in the soil under a thin shading of palm-leaves, and the foliage syringed morning and evening; this will induce them to make strong healthy growth, and form nice plants for the winter. Crotons must never be allowed to remain dry at the roots, or they will lose their leaves, and tall leggy plants will be the result. When this occurs the tops should be taken off in the autumn and struck in pure sand, the leaves being tied up to prevent them from flagging, and the cuttings planted singly in small pots, and put into a close glass frame or under bell-glasses, until they have rooted, the moisture being wiped off the inside of the glass every morning, and air admitted by slightly raising the glass as the cuttings begin to root.

The old stumps may then be cut down to within a foot from the pot; plunge in a half shady position, and if not watered too often, they will make nice bushy plants in the following spring.

Crotons cultivated in this country, if subjected to the strong damp heat of a hot house, as in Europe, quickly lose their lower leaves, while cold treatment, will be found to produce dwarf bushy plants and well-marked foliage.

The following are a few varieties worth growing:—
C. Andreanus, Elegantissimus, Chantrierii, Interruptus aureus, Lady Zetland, Langii, Mutabilis, Pictus, Queen Victoria, Sunbeam, Williamsii, Superba.

Datura alba.—A handsome, well-known plant with large white pendant flowers, and rough hairy leaves. It can be grown either as a standard or a bush. The flowers, which are borne in great profusion, have a strong scent, and are said to produce sleep by their narcotic properties. The plant should be well supplied with manure-water when the buds appear, and pruned back in the autumn.

D. suaveolens is also grown as a garden shrub; while *D. Stramonium, D. Metel,* and *D. fastuosa* grow wild in the country.

Dracæna ferrea.—A very ornamental variety, usually growing in clumps to the height of 3ft. to 5ft. It grows freely in rich sandy soil, in the open ground, but it should be planted in a position where its leaves are not broken by the wind. Many beautiful red and variegated varieties are also grown in the plant-houses, and succeed in a compost of leaf soil and sand; weak manure-water should occassionally be given during the summer if the plants have plenty of roots. All varieties can be propagated from cuttings, or from the thick base of the root-stock, which will sometimes grow out of the bottom of the pot. This may be cut off and planted.

The following varieties are suitable for cultivation in glasshouses:

D. Bausei, Cardeii, Eburnea, Gigantea, Imperialis, Nobilis, Norwoodensis, Prince Manouk Bey, Splendida, Superba, Youngii, Madam Salvago, Gladstonii.

Hibiscus mutabilis — The Changeable Rose. — A tall, bushy, deciduous shrub, with hairy palmate leaves, and blooming in the autumn with large rose-like flowers, which are white in the morning, pink at noon, and of a deep rosey tint at evening. It will grow on very poor soil, and is easily propagated by cuttings, when the tops are pruned back in February.

H. schizopetala.—A variety with small leaves and hanging flowers, with fringed petals. Lately introduced into the Ramleh Gardens.

H. rosa-sinensis. — A handsome garden shrub. For description, see page 14.

H. syriacus—The Syrian Hibiscus.—There are seven varieties of this species, but the *albus plenus*, or double white, is the one commonly met with. Its stems are clothed in July and August with white sessile flowers, known by the Arabs as "Gotton." The plants are sometimes used for hedges.

Many double varieties, ranging in colour from a dark red, to a beautiful orange, are also grown from cuttings, and make handsome flowering shrubs, which require pruning annually to keep them bushy.

Pittosporum undulatum. — A well-known shrub, common in the gardens, and without doubt the best plant for a windy position. It can be used for a similar purpose as *Euonymus japonicus*, which is so commonly planted on the south coast of England.

P. variegatum. — This is a handsome silver-coloured variety, making it a pretty shrub for borders or a lawn, but is less hardy than the former species.

Both are propagated by cuttings made of the half-ripened wood in the spring or the autumn.

Rhododendrons, Camellias, Azaleas, and Gardenias, bloom fairly well the first or second year after being brought into the country, but they usually deteriorate with the heat and the lime in the water, and unless fresh importations are constantly being made, these plants can scarcely be considered a success in Egypt.

New Holland plants, such as *Banksia, Callistemon, Hakea, Melaleuca, Metrosideros*, &c., do not succeed in the salt and limey soil of Alexandria, but many of the Australian trees grow to perfection in Cairo.

The following is a list of twenty-six varieties of shrubs that are established in the country, and can be recommended for their foliage or flowers :—

Acokanthera spectabilis
Calotropis procera
Carica Papaya
Deutzia scabra
Eranthemum nervosa
Eugenia australis
E. Jambosa
Jacobinia magnifica
Lawsonia alba
Lagerströmia indica
Meryta angustifolia
Murraya exotica
Melaleuca ericifolia

Melianthus comosus
Nandina domestica
Nyctanthes Arbor-tristis
Ochrosia elliptica
Phyllanthus angustifolius
Rondeletia speciosa
R. variegata
R. versicolor
Russelia juncea
Sambuscus nigra
Sparmannia africana
Spiræa prunifolia
Syringa persica

CHAPTER VI.

CLIMBING PLANTS.

EGYPT is very rich in the number of its climbers, for not only will those that one often sees growing in the open air in England thrive here, but many of those which generally require glasshouses and heat in Europe, do perfectly well in Egyptian gardens, providing that the proper position as to sun and shelter, is thought of when planting.

Two important points should always be remembered when planting climbers, position and pruning, while arrangement of colour, and time of flowering should also be taken into consideration.

In many gardens a winter and spring effect is perhaps more desirable than a summer one. Care should therefore be taken that a prominent position is not occupied by a plant which loses its leaves in winter, or that its place is one so exposed that at the time of flowering the bloom is destroyed by the cold spring winds. The *Bougainvillea*, which is a brilliant mass of flowers during the winter months, is a good example of the necessity of protection against the wind.

Climbers should not be planted under the shade or near the roots of large trees (except, of course, when required to climb up their trunk, as in the case of ivy, &c.), for in such a position it is unreasonable to expect good results. It would be advisable, also, to learn beforehand the character of the plant, and to consider whether the space allotted to it is sufficient for its development; and lastly, it is necessary that a bed should be prepared with at least a metre of rich soil in order that the young plant may make a good start.

TRAINING AND PRUNING.

The foundation of the future plant should, if possible, be made when the specimen is still young, and the shoots should be spread out evenly, and trained in a fan-shaped or upright form as the position may suggest. Shoots should never be allowed to cross one another, and only those should be permitted to remain that the space can accommodate, all others being cut away. Flowers can only be obtained from those shoots that are exposed and ripened by the sun; hence care should be taken that each shoot is at an equal distance from the other, and no crossing or recrossing of the branches should be allowed, as otherwise the plant will present a confused mass, and only the upper and exposed parts will flower.

Old shoots should be replaced by young ones, and in some species, the red *Tecoma capensis*, for example, the climber may be improved by having each alternate shoot pruned half-way back, so that the flowers may be distributed equally over the whole plant, and not simply blossoms in a mass at the top, as is so often seen at present. All climbers should be pruned at least once a year, the usual time being after flowering.

It will be noticed that climbers on a wall invariably suffer from the strong, latent heat, and therefore if they can be trained over lattice-work or on bamboo screens placed about 1ft. from the wall, so that the air can pass at the back, better results will be obtained.

The following is a selection of some of the most suitable for planting :—

Antigonon leptopus.—A lovely deciduous climber common in the gardens of Alexandria. It has small heart-shaped leaves, and blooms in October with drooping masses of small pink flowers. It propagates easily from seed, or small plants in pots can be obtained from the local nurserymen. A charming plant for climbing over the entrance to a verandah, summer-house, or over garden arches.

A. insignis.—Similar to the above-named variety in habit, but has flowers of a lighter colour. This variety has not yet been introduced into the country.

Aristolochia brasiliensis — Pelican Flower.—A tall climber, with large cordate leaves, and curious flowers of an indescribable form, greyish in colour, with small brown spots, and of a disagreeable odour. The plant is propagated from seed, and grows well in Cairo.

A. elegans. — A handsome climber for a greenhouse. The plant is of pendant nature, with smooth light green leaves and hanging flowers of a shell-like form, which are entirely free from the objectionable odour peculiar to the genus.

A. gigas.—This has one of the largest flowers in the vegetable kingdom, and many other varieties might be introduced with success into this country.

Beaumontia grandiflora.—This beautiful climber is worthy of a place in all sheltered gardens; though, unfortunately, it is only rarely met with in Cairo. It is, perhaps, seen at its best when climbing the stem of a tall palm-tree, to which it firmly attaches itself by means of its rope-like tendrils. It has rough, oval leaves, and corymbs of large trumpet-shaped flowers resembling white lilies. Young plants can be obtained from seed, or propagated by layers and cuttings.

Buddleia madagascariensis.—Commonly met with in the gardens here. It is of rampant growth, thriving in almost any soil and position. A suitable plant for covering a wall on the outskirts of a garden. It has long, lanceolate leaves, that are white on the under surface, and is one of the first of the climbing plants to bloom—flowering early in February—with drooping clusters of small flowers of a pale orange colour.

Easily propagated by cuttings. It should be cut in closely after flowering, or it will become very straggling.

Bougainvillea spectabilis.—A strong-growing climber, with thorny stems and hairy foliage, requiring a high wall with a south aspect, as it blooms in the months when the strong sea winds are prevalent. Its masses of majenta-coloured bracts, often extending over the entire plant, makes it an effective object during the winter months. It requires to be pruned after flowering, and all tall shoots should be cut back during the season. Propagated usually by layers. It looks best when grown in a mass, as its majenta-coloured bracts are difficult to blend with other garden flowers.

B. glabra.—This variety is seldom grown here. It has smooth, shining leaves, but the bracts are of a light mauve colour, and are not so numerous as in the former. A dark, brick-red variety, with smaller bracts than the others, is often seen in the gardens of Alexandria.

Cryptostegia grandiflora.—A rather unmanageable climber, possibly of more value as a rubber plant than for the garden. It has handsome luxuriant foliage, large purple flowers, of a bell-shaped form, and a curious angular fruit which always grows in pairs. It flowers in the autumn and winter months, and is a suitable plant for covering an outhouse. Propagated in February by cuttings.

Clitoria ternatea—Mussel-Shell Creeper. — Native of the West Indies. A pretty deciduous climbing plant easily raised from seed; leaves pinnate, with two or three pairs of ovate leaflets, and handsome pea-shaped flowers of a deep indigo blue. The plant grows to the height of about 6ft. It can be cultivated in pots, trained on bamboo sticks, or over lattice-work. There are two or three other varieties, both single and with double flowers, also a pure white variety; but these are somewhat rare.

Clematis Jackmanii.—An elegant climber, growing well in the gardens here, but rarely met with. It has large, four-petalled flowers of a deep violet colour, blooming profusely in April and May. A suitable plant to cover arches or a summer-house.

C. Princess of Wales bears a large star-like flower of a deep heliotrope colour. It is a shy growing plant, and is not so hardy as the former.

Other varieties, such as *C. coccinea*, Duchess of Edinburgh, and *montana* are occasionally seen, and many others would no doubt succeed if they were trained on lattice-work in sheltered spots.

Hedera Helix — The Ivy. — This old English favourite is too well known to need description. It may be utilised with advantage for training up the stems of palms and large trees, where it succeeds well in the half-shade, imparting a green, furnished appearance. It will also grow on rails or trellis work; but, if close to the wall, its leaves invariably become yellow, and covered with a black scale insect. A very pretty small-leaved variety, of a slower growth, can be used for a similar purpose, and many others might be introduced.

Hoya carnosa—Wax Plant.—A familiar plant, and the best of the species, thriving on rockeries in a half-shady position. The roots delight in a moist, light soil, leaf mould and broken brick rubbish suiting it admirably. Its thick, glossy leaves, and umbels of wax-like flowers, are known to many. The plant is propagated by layers and cuttings.

Other species with inconspicuous flowers are sometimes seen.

Ipomœa—*Bona-nox*—Good-Night Plant.—A convolvulus-like climber of somewhat straggling growth, often seen climbing over verandahs, for which it is especially suitable on account of its large, trumpet-shaped flowers, which open at night, throwing out a pleasant perfume.

Many other varieties of more or less value are also grown, the commonest of which is *I. cairica*, a native of Cairo, known to the Arabs as *Sitt-el-hosn*.

Jasminum grandiflora.—One of the best of the many varieties of Jasmine, with graceful pinnate foliage, and leaflets less than an inch long, and cymes of white fragrant flowers. The plant is of a somewhat sprawling nature, and is rather troublesome to keep in order, its trailing stems emitting roots wherever they touch the ground.

J. officinale.—An old familiar favourite of our English gardens, also *J. confusum*.

J. Sambac—The Arabian Jasmine.—A thick, close-growing variety, with numerous box-like leaves and small white flowers with long stamens.

The Great Double Tuscan Jasmine, known to the Arabs as *Foull*.—It has strong-scented flowers like small white roses, but it is more of a shrub than a climber. The plant requires a rich sandy soil and a half-shady position.

J. revolutum.—A winter flowering yellow variety.

J. humile and **J. fruticans,** also yellow varieties, are occasionally grown.

Lonicera chinensis — Honeysuckle. — A favourite plant in Egyptian gardens, filling the air at evening with a delicious lemon-like perfume. It is of rapid growth when once established, but the branches, which twine and interlace with one another, makes frequent pruning necessary to prevent it from becoming an entangled mass. It is capable of growing on very poor soil, but is too well known to need describing. *L. sempervirens*, var. *minor*, a deep red variety, should also be grown.

Momordica balsamina.—A pretty annual gourd, growing easily from seed, which should be sown in March, and bearing in August an orange coloured fruit resembling a ridge cucumber. This bursts when ripe and exposes a mass of red pulp, in which numerous seeds are embedded. The pulp, which is of a sweet sickly flavour, is sometimes eaten by the natives. It is also bottled in olive oil, and after being exposed to the sun, it is used for healing cuts and wounds, with, it is said, excellent results.

Lagenaria vulgaris (The Bottle Gourd) and **Luffa cylindrica** (The Luffa), which are used as flesh sponges, are also cultivated from seed, in the same manner as the former plant, and are usually trained over summer-houses or trellis-work, that the large pendant fruits may be exposed.

Periploca lævigata.—A plant of the same family as the *Cryptostegia* (*Asclepiadaceæ*). It has lanceolate leaves, and clusters of inconspicuous, star-shaped flowers, bearing two long finger-shaped seeds joined together. This plant is often found growing in the gardens owned by Greeks.

Quisqualis indica.—An Indian plant of medicinal properties, requiring strong rails or trellis work for its support. It bears during the summer months, profuse clusters of fragrant white flowers of a jasmine-like form, which turn the following day to a deep red. This mixture of the two different coloured flowers gives the plant a very singular appearance. Propagated by layers and cuttings.

Rhyncospermum jasminoides.—Native of China. A charming plant, with slender stems, growing to the height of about six to eight feet, bearing in April masses of very fragrant pure white flowers, which often clothe the entire plant with blossom. Next to the *Antigonon leptopus* it is probably one of the best of our small flowering climbers.

Stephanotis floribunda.—Native of Madagascar. A familiar plant in our hot-houses in England. Here in Egypt, with a continual sunshine, it is one of the handsomest climbing plants. Its clusters of pure white flowers fill the gardens with perfume in the early June mornings. The thick, fleshy nature of the leaves, enables the plant to remain for a considerable time without water, although the damp climate of Alexandria is said to be more conducive to its flowering properties, than the dryer climate of Cairo, where it should be frequently syringed.

It is usually propagated by layers, but as it blooms on the young wood it should only be thinned of any superfluous growth, and sufficient room allowed to prevent it becoming a crowded mass.

Solandra grandiflora.—A strong-growing climber, with thick, succulent shoots, and large, smooth leaves, of an oval-lanceolate form ; bearing, in April and May, an erect, horn-shaped flower, with a white overlapping rim, which turns the next day into a creamy-yellow. It is easily propagated by cuttings.

Solanum Seaforthiana.—A pretty climber for a glasshouse. It flowers in the autumn, with hanging clusters of small blue flowers.

S. Wedlandii.—A handsome species, worthy of introduction for the conservatory.

Tecoma capensis.—An evergreen plant, common in the gardens here. It can be either trained as a climber, or cut back into a shrub. It also makes a pretty pot plant for the verandah. It has dark green pinnate leaves, and masses of red labiate flowers, will grow in any soil and position. It is a valuable addition to a collection, as it flowers throughout the year.

T. radicans. — Like the former this is a well-known plant, but its large pinnate leaves are deciduous. It blooms in the summer months with terminal corymbs of tubular orange-coloured flowers.

T. jasminoides.—An Australian variety, with pink funnel-shaped flowers, is also worth growing.

These can be propagated by layering the lower shoots, or by cuttings.

To this list may be added many other climbing plants, worthy of a place in every collection of climbers. Such as *Ampelopsis Veitchii*. *Phaseolus Caracalla* (the Snail Plant).—*Dolichos sinensis*, *Wistaria polystachya*, *Boussingaultia baselloides*, whose drooping racemes of small white flowers, smell like new-mowed hay. *Lantana nivea*, and its many different coloured varieties, which grow well on a hot wall, but require constant clipping.

Passiflora edule and **P. quadrangularis,** both of which are handsome climbers, and furnish an edible fruit if artificially fertilised. The latter is known under the name the Granadilla.

P. cœrulea and **P. Constance Elliot** are also well known varieties, though the foliage and flowers are not so bold in character as the two former. The passion flowers all propagate easily by layers, *i.e.*, the lower stems having a cut made at a joint, and then peg down in the soil, or in pots until they have rooted, after which they can be disconnected from the old plants.

Pleroma macrantha.—Native of Brazil. A pretty climber, with hairy pinnate leaves, and large violet-purple flowers. The plant is suitable for a greenhouse or a sheltered position.

CHAPTER VII.

ROSES.

OF all flowers roses are probably the most popular, and it is therefore not surprising, to find that in every garden a large space is devoted to their cultivation. But the great fault to be found in this country is the lack of variety. Garden after garden may be visited with the result that while blooms are to be seen in abundance, the kinds are the same in all. This cannot be attributed to a lack of enterprise on the part of the owners, for much money has been spent and many beautiful plants have been imported from Europe; but the selection has seldom been a judicious one, and a choice has often been made of roses which, while thriving well in Europe, are unsuitable for the Egyptian climate.

Such roses are sometimes a partial success the first season, but many are lacking in size and in the colour of their blooms, and those that do not die off during the first summer, usually exhaust themselves by throwing up long shoots, at a time when they ought to be at rest; consequently they deteriorate, and are ultimately lost.

Several of the *Hybrid Perpetuals* which hold so high a place in England, are unsuitable for Egypt, and preference should be given rather to roses of a more tropical nature, such as Teas, Noisettes, Bourbons, China and their hybrids, which usually flower more or less the whole year round, while those of a more temperate nature, instances of which we have in the Cabbage rose, *Nero*, and the majority of the European roses, bloom only in April and May.

Roses delight in a rich heavy soil, and, consequently, if the ground is of a poor sandy nature, the beds should be deeply dug, and a good dressing of Nile mud and manure applied. Permanent beds should have an annual dressing of well-rotted manure, that has decomposed in a damp state, not the dry dusty compost that has lost most of its fertilising properties, one so often sees used by the native gardeners. This dressing should be given about the middle of February, and a light mulching at the rise of the Nile in August, will be found to be of great help to the autumn blooms. For this purpose it will be best to place the fresh manure in a hole, where it can be kept moist, until it becomes decomposed and develops its full properties.

In forming rose beds care should be taken that they are within reach of a good supply of water, as they will require watering frequently during the months of March, April, and May, after which less water should be given when the spring flush of blossoms are over, to prevent weak, yellow growth; but the supply may again be increased at the rise of the Nile. Climbing roses will invariably be found to succeed better when trained on trellis-work, or on arches, than when planted close against a wall, where they usually become sickly and burnt from the strong latent heat the wall gives out.

Pruning.—The end of January, or the beginning of February, is usually the best time for pruning roses. But with regard to La France variety, opinions differ as to whether it is better to prune the whole garden at once, or to divide it into sections, and to prune a portion, each following week or ten days.

If done in sections, pruning should be begun in the most sheltered part of the garden, as early as the second week in January; but it ought never to be attempted unless the stock of plants is a large one, and unless the owner is willing to run the risk of having some of the shoots nipped off by the strong salt winds. In gardens where quantities are grown, the risk is well worth running, as it can matter little, whether some of the earlier blooms are checked; and should, on the contrary, the season be mild, the young shoots will push ahead and give blooms earlier than if the whole pruning were done at the usual time mentioned above. The first blooms of La France are always better in form, and in colour than those that appear later.

Climbing roses should have their long shoots cut back a foot, and all crowded, and weakly stems thinned out. Cuttings may be planted in February.

Budding.—This can be done either in the spring about the end of February or after the rise of the Nile in the autumn, but the spring budding gives a greater percentage of "takes" with roses that are grown on light soil, and are considered stronger than those that are budded in the autumn; though the latter time would perhaps be best for roses grown in the heavy soil of Cairo, as the wood is usually too hard for the budding to be a success in the spring.

The Tunisian Briars, known as *Magrubies*, and the Gloire de Dijon stock, are the most suitable for budding purposes, though buds often succeed on other stocks. Many varieties also succeed from cuttings which may be made at the same time.

The Rose Beetle *(Cetonia aurata)* is a very troublesome insect to the early spring blooms, eating its way into the heart of the flower, and doing great damage to the petals. The only way to remedy this evil is to go over the plants in the early morning before the dew is off, and again when the sun is up, and pick off the insects and destroy them.

The following is a list of roses already cultivated in the country, and others that might be introduced with a fair hope of success.

Climbing Roses.

Aimée Vibert.—A pretty white Noisette, with small flowers borne in crowded trusses, having a musk-like odour, and becoming tinged with red as they fade. This variety is, perhaps, a typical Noisette, the origin of which tribe was derived from a cross between the Musk Rose and a common China. It is an excellent plant for covering pillars, and, if required for that purpose, it should be reduced to about three stems. These should have their tops shortened a foot, and the side shoots pruned back in November, and then they should be unfastened and laid at full length along the ground for two months; this will induce them to produce flowering shoots along their whole length instead of only a few bunches at the top as one so often sees. In the early spring they should be carefully raised, and tied into position. Propagated by budding, layers, and cuttings.

Fortune's Yellow Rose.—A rambling, quick-growing variety with slender stems and small, fawn-coloured flowers. The blooms which are borne on the upper part of the stems will be destroyed if the shoots are pruned. Therefore an occasional *thinning* of dead or weak wood is necessary. Propagated by layers.

Gloire de Dijon.—One of the finest of the tea roses, and also one of the strongest in growth, too common to need description. The colour varies from the palest yellow, to reddish orange in some of the flowers. It is propagated by budding on briars.

Maréchal Niel. — A lovely yellow, tea-scented, Noisette, thoroughly established in Egyptian gardens, flowering abundantly about the middle of March, and continuing more or less throughout the summer. It is of vigorous growth when trained on trellis-work or where the air can pass through, and is considered to do best when budded on the Gloire de Dijon stock.

Marie Henriette.—A large, handsome pink rose, occasionally met with in the Ramleh gardens. Propagated by budding or cuttings; but budding is the best.

Rosa ternata.—A Banksian rose, with glossy ternate leaves, and white flowers.

Rosa Noisettiana Fillenberg—*Stambouli.*—A deep red native variety, of which there are several forms, some having small flowers with incurved petals, while in others the flowers are larger, of a lighter colour, and petals fringed. The smaller variety, which is more common, can be either grown as a climber for covering stems of date palms, &c., or as a bush. Beds of the latter form, are valuable for a winter effect, the flowers contrasting pleasingly with their dark green foliage. The cut flowers are also useful for the decoration of the dinner-table. Propagated by cuttings.

White Banksian.—A climbing rose, with thornless twigs, and long narrow leaves. It flowers in the spring with bunches of small white blooms. Propagated by layers.

Yellow Banksian.—An exceedingly pretty dwarf climber, flowering in clusters.

William Allen Richardson.—A favourite Noisette, with copper-coloured flowers in the spring, and yellow in the autumn. They are seen at their best when in the bud. It is a shy grower, and does not succeed in every garden, but could no doubt be improved by working it on vigorous stocks.

HALF-CLIMBERS.

Boule de Neige—The Snowball or Camellia Rose.—One of the few Hybrid Perpetuals that succeed fairly well in this country. It is propagated by cuttings.

Homer.—A tea-scented variety, common in the gardens, making a fine object either as a bush, standard, or specimen plant on a lawn. It has healthy, dark green foliage, with red tips, which, unlike the majority of roses grown in Egypt, the leaves remain in a fresh condition throughout the year.

The flowers are white, tinged with red, but are effective only in the bud. It blooms continually at all seasons, and, with the exception of an annual thinning, and cutting into shape, the plant requires little or no pruning. Propagated by budding.

Safrano.—A very effective tea rose. It can be grown either as a half-climber, or a large bush. It has glossy, dark green foliage, and young shoots of a coppery tint. The flowers are of a deep fawn colour, beautiful both in the bud, and also when they first open. It can be propagated by budding, but is a somewhat rare plant in the gardens here.

DWARF ROSES.

Bourbon Rose—Little Gem.—The smallest of all roses, forming a very dwarf bush, and bearing a number of small pink flowers. The plant is often used for filling in the centres of large flower-beds. Propagated by taking off the stems with a heel, which quickly root in February.

Caprice.—A charming dwarf rose, with flowers of a yellow pink and orange tinge, which often changes. They soon fade, however, when cut.

La France.—A Hybrid Perpetual. The commonest rose in the country, bearing a large pink bloom, which very soon deteriorates in colour and texture with the age of the plant or the quality of the soil. In rich, heavy soil it will last several years, but in soil of a poor, sandy nature, it will be found advisable to lift the plants annually in February, and dig a good dressing of manure into the beds, and, before replanting, prune the stems to three or four eyes, and cut the thick tap-root back; this will induce the plants to make fibrous-feeding roots, which will add to the substance and colour of the flowers. They are easily propagated by cuttings, and some growers dispense with the old plants, after the third or fourth year.

Perle des Jardins.—A dwarf tea rose, with brown tinged foliage and large yellow blooms, borne on long upright red stems. A common rose in Cairo. Is propagated by budding, layers, or cuttings.

Rosa rugosa.—A very coarse, pink rose, with thickly-thorned stems, and rough leaves, usually grown in this country as a dwarf hedge for the kitchen garden. The petals are used for making a preserve. It is propagated easily from suckers, and when once established it will quickly spread, and is even sometimes difficult to keep in its place.

Rosa centifolia—The Green Rose.—This variety has the same habit as the ordinary dwarf rose, but the flowers and buds are of a deep green colour, perhaps more curious than beautiful.

Souvenir d'un Ami.—A first-class, tea-scented rose, with pink-shaded flowers, which are very beautiful in their drooping, half-expanded form. Propagated by budding.

Souvenir de la Malmaison.—A very popular and well-known variety of the Bourbon tribe. The flowers are whitish, with a rosy tinge in the centre, but the foliage is often attacked by mildew. Many opinions are expressed as to the proper method of culture and propagation: some prefer layering the large shoots, and dispensing with the old plants after the third or fourth year; while others recommend grafting on the *Stambouli*, and other stocks; but this must be governed by the locality, for, where one method does not succeed the other should be tried. Old plants, by having the soil partly removed, some of their thick tap-roots cut off, and a dressing of rich soil and manure applied in the early spring, will often continue to bloom for a number of years. When transplanting, this variety should be removed with a ball of earth, as the leaves always remain, and during the flowering season old flowers should be picked off, and withhold water after May, as recommended with La France.

The following is a list of a few varieties belonging to the tribes that are already established in the country, and which might be tried with a reasonable chance of success in Egypt.

TEA-SCENTED ROSES.

White. — Climbing Devoniensis, Madam Bravy, Madam Willermoz, Niphetos, Rubens, Triomphe de Guillot.

Yellow. — Bouton d'Or, Isabella Sprunt, Madam Falcot, Louise de Savoy, Narcisse, Pactolus.

Fawn and Salmon. — August Oger, Madam St. Joseph, and Triomphe de Luxemburg.

Flesh and Blush. — Catherine Mermet, Cheshunt Hybrid, Maréchal Bugeaud, and Sombreuil.

Red.—Rosa-indica sanguinea.

NOISETTE.

White. — Jeanne d'Arc, Maria Massot, White Noisette.

Red.—Sir Walter Scott, Lady Buller, Bridesmaid.

Yellow.—Earl of Eldon, Jane Hardy, Claudia Augustin, Magarita.

TEA-SCENTED NOISETTE.

Yellow.—Cloth of Gold, Isabella Gray, Triomphe de Rennes.

CHAPTER VIII.

CONSERVATORIES AND GLASS-HOUSES.

CONSERVATORIES and greenhouses for the cultivation of choice tropical plants, or for the protection of those of a delicate nature during the winter months, have been, until within the last few years, comparatively unknown in the smaller gardens of this country. Many large plant houses, however, exist in the first-class gardens in Cairo, though, perhaps, their beauty is not so great to-day as it was formerly. In some cases these houses have been purchased and re-erected by private individuals in different parts of the country, and others, following their example, have brought iron houses from Europe, or built them in a more temporary manner with wood; but, unfortunately, in most cases the houses present anything but a pleasing effect, owing to the inability of the native gardeners to manage them, and who overshade, and neglect to ventilate and clean them, besides often filling them with weak and common plants which would do much better in the open air.

In the more modern gardens a small number of well-kept glass-houses have lately sprung up, and a few wealthy amateurs have spared no trouble or expense, in getting together a choice collection of tropical foliage plants, ferns, orchids, tender climbers, and others.

There are two kinds of glass-houses in Egypt: the first where the health of the plants are considered.

These are usually low, span-roofed structures, known in England as "pits," having their staging or platforms level with the ground, and the slanting roof rising a few feet above. Ample provision should be made for ventilation, by having a door at each end, and the top and side lights should be made to open, while a tank may be let into the stage in the middle of the house for the supply of water. The outside could be improved by having a lawn and flower-beds surrounding the house, and the sloping entrance tastefully decorated with rock-work, or a large specimen plant each side. Shade can be obtained by rolling-blinds, which should be drawn up when the sun is not shining, and the roof may easily be cleaned from the dust by an occasional washing with the hose. The advantage of a low roof is that it enables the plants to be kept near the glass, which promotes strong robust growth, and prevents sickly drawn plants.

A very convenient staging which will impart a tasteful appearance to the house might also be built of rocks, which, if obtained from the coast, should be previously soaked in fresh water to free them from the salt, and then planted with Ferns, *Begonia Rex*, *Fittonias*, *Lycopodiums*, *Marantas*, *Sanchezia nobilis*, *Peperomias*, &c., which often grow like weeds when planted out, and the owner would thus have a bank of foliage reaching from the ground to the roof instead of an open, unsightly staging. From the roof could be hung Orchids, *Nepenthes* (pitcher plants), *Platycerium alcicorne* (stag's horn fern), *Ophioglossum pendulum* (ribbon fern), *Eschemanthus*, and other suitable plants. Houses of this kind are warm in winter and protected, they are also cool, and retain the moisture in the summer, by being low in the ground, besides costing about half the amount of a tall, elevated conservatory.

The next class of plant houses, are the tall conservatories which are generally raised a step or two above the ground, or at least level with it; these may be either lean-to, a tall span, or arched roof. Houses of this kind are very decorative in themselves, and are also suitable for tall palms, climbers, and as a show house; but the former class are the best for dwarf plants.

The site for a glass-house, should be a sheltered spot, protected from the wind, but not shaded by trees. From east to west is perhaps, the most suitable, as these are points which escape the wind, and are exposed to the sun during the winter. Iron framing, is of course, more durable than wood, as the latter deteriorates with the constant heat of the sun and the moisture inside.

Shading should be either of a light permanent nature, with just sufficient thickness to break the fierce glare of the sun, or in the form of roller blinds.

Furnishing, must of course, rest with the owner; but where the house is wide enough, a central stage, with a convenient path on each side, should be made for tall plants, while the smaller ones might be arranged round the side stages, and provision made for carrying off the water from the paths.

The water supply may be obtained from a pool with cemented sides, fed by an artificial waterfall of rock-work, which could be tastefuly planted with ferns, mosses, and other suitable rock-plants.

If the stages are built with rock-work, a space should be left at each support, which should be filled with rich sandy soil, for planting out climbers that may be trained up the woodwork, and allowed to hang in festoons from the roof. These should be limited to *two* main stems only, so that after pruning they may be taken down and cleaned, and the paint and inside glass washed.

Baskets and hanging plants may also be introduced and hung at intervals along the roof.

Considerable taste and judgment, ought to be exercised in arranging the plants, the tallest behind, and the shorter in front, placing the red and variegated foliaged plants at intervals, so as to harmonise and create a pleasing effect, and the edges of the stages should be

F

lined with maidenhair and other ferns, hanging-grass of *Panicun variegatum*, and other suitable plants.

Small shingle, obtained from the seashore, or the round marble-like quartz from the Makattam Hills and other places near Cairo, are the best materials for the tops of the stages, which should be previously filled in below with broken stone, brick, or any rough material, and rich, sandy soil filled in amongst the rocks. Sand should never be used for the surface of the stages, as it has a tendency to stop up the bottom of the pots, causing them to become water-logged.

The rocks obtained from the coast near Mandarah, or Mex, are suitable when the salt has been washed out of them, but the soft yellow sandstone from the desert is preferable, as the plants cling to it readily, and many things that will not grow, or only exist in pots, such as *Marantas*, thrive amazingly when planted out in rockwork where no cement has been used, and many of the choice varieties of ferns spore and grow readily in it.

MANAGEMENT.

Ventilating, or giving air, is an important item in the management of a glass-house, and with the exception of severe winter gales, or a Khamseen, an inch or two should be left open at the top night and day throughout the year. During the winter, the houses should be open daily on the side where the wind is not blowing, and should be closed in the afternoon half an hour before the sun goes off them, the houses having previously been "damped down" by having water sprinkled on the paths and stages. This will cause a moist growing heat to rise, and will also keep the temperature a degree or two higher than the outside through the night.

Damping the paths, either with a small hose or a water-can, should be done several times during the day, and always in the morning and afternoon when opening and shutting the houses. Syringing occasionally

has a beneficial effect on palms, but filtered water
should be used, or the foliage will be covered with
a white limey sediment.

Mixed collections of plants, and especially Maidenhair ferns (*Adiantums*), should not be heavily syringed,
as the water causes the young fronds to rot, and
therefore, where these are present, it will be advisable
to only damp the stages with a small can, which might
be done three or four times a day, according to the
season, though a fine vapour-like syringing about eight
o'clock in the moring of a very hot day, will prove
beneficial and help to keep down insect pests, which a
dry atmosphere promotes.

If shading cannot be carried out by means of
roller-blinds, as previously mentioned, a thin gauze
tacked over a light wooden framework about May
will be found to answer the purpose, or the glass
can be painted with a mixture of milk and whiten-
ing, care being taken that the shading is not too
heavy, or sickly plants will be the consequence.

As the plants naturally turn their foliage towards
the light, it will be advisable to re-arrange them
every month or six weeks, so as to have the better
side seen from the paths.

Cleaning.—For this purpose the plants should be
taken out in batches, according to the size of the
house, and thoroughly overhauled, having their foliage
cleaned from dirt and insects, the pots washed, and
the drainage examined, and any weeds or scum on
the soil removed. The shingle should also be re-
arranged, dirt on the woodwork or glass washed off,
and the plants returned to their places before the
evening. If this system, of a thorough cleaning
throughout the house is carefully carried out, few
dirty plants will ever be seen, and the house will
always present a clean and orderly appearance.

Potting.—This is an operation that should be carried
out with considerable care and forethought, as upon it
the beauty and health of the plant greatly depend.

In the management of a collection of choice tropical plants, something should be known of the habitat and natural conditions, under which the plants usually grow. The few following remarks, gained from practical experience in the cultivation of choice stove plants and orchids in Egypt, may I hope be of some assistance, to the amateur who is desirous of adding to his collection.

Always use clean pots; new ones should be first soaked in water before using. It will also be found best after potting to have the old pots washed, and stacked away according to their sizes, so as to be ready for further use. Broken ones should also be washed, as they can be used for drainage, or, as it is technically called, "crocking the pots." New pots usually arrive from the pottery with the hole at the bottom much too small for draining purposes. These should be made larger, and a curved piece of pottery placed over the hole, before the smaller pieces are put in. The native method, however, is to put one stone at the bottom, and this invariably causes the soil to become sour and water-logged, by stopping up the drainage, so that the plant sickens and dies.

Anthuriums, Philodendrons, and all Aroids should have their pots half-filled with drainage, and a little moss or dried grass, placed over the top of the crocks to prevent the soil from washing down amongst them. Rich, sandy-leaf soil, mixed with chopped sphagnum moss, and a little charcoal, or pottery dust, is an excellent compost for them. The Sphagnum moss can be obtained dry in sacks, from any European nurseryman, and should be steeped in water, and then squeezed out, and cut finely with a pair of scissors. No amateur who cultivates stove plants in Egypt, should be without this moss.

When potting Aroids, which are one of the best class of plants for an Egyptian glasshouse, see that the crowns are above the rim of the pot, and finish off with Sphagnum moss, so that the plants are raised on a mound of moss, and fresh supply ought to be added as the surface roots appear. Plenty of water should be given at all times, and weak cow-manure water twice weekly during the summer months will be of service. Artificial manures are also of great benefit when judiciously used.

Manure water poured on the paths late in the afternoon, after the houses have been closed, and the heat risen, will cause the ammonia to rise and promote dark healthy foliage.

Terrestrial Orchids.—Plants of this tribe that are cultivated in pots, pans, and hanging baskets, may be potted in the same compost as the former. The compost, however, if used, should be as coarse as possible, and pieces of charcoal about the size of pigeon-eggs and plenty of chopped Sphagnum moss added.

Celestial Orchids.—These may be placed on pieces of virgin cork, or charred wood with a little moss, and firmly secured by copper wire. In this way the plants will soon form roots which attach themselves to the cork and wood, as long as the atmosphere is kept moist, and they are not allowed to suffer from dryness. Cool orchids, as *Odontoglossum*, &c., are in most cases a failure.

Adiantums.—These ferns thrive well in a compost of leaf-soil, sand, and loam, with a little broken pottery or brick added. They require copious supplies of water during the summer, and should never be allowed to get dry. In growing large specimens, particularly *A. Farleyense* (the handsomest of all Maidenhairs), I have found it a good practice when potting to place a small inverted pot over the drainage-hole inside the larger one, and fill the space between the two pots with small pieces of pottery and a thin layer of dry cow-manure placed over this, using a little heavier compost and leaving an inch or two below the rim of the pot to hold the water, but keep the fronds as dry as possible. All Maidenhairs in the autumn, when their fronds turn brown, should have them cut off level with the crown of the plant, which should be kept drier until the new fronds appear in February, when they may be shaken out and re-potted, the larger plants being divided if the stock is to be increased. Young plants will often be found growing on the rockwork, these can also be potted. All varieties will be benefited by having soot and cow-manure water applied alternately twice a week during the summer.

Potting Soil.—For ordinary potting purposes the turf cut from damp places and stacked grass downwards in a disused part of the garden, with alternate layers of cow-manure and a little sand and charcoal-dust between, will in time form an excellent potting compost. The whole should be well watered, and allowed to remain for six months or longer to become fallow, after which it may be cut down with the fas or spade, and a little sand added.

For palms and the ordinary garden plants this will be found to be the best soil obtainable in the country for ordinary potting purposes.

Leaf-soil.—This can be obtained by placing all the garden refuse into a large hole, filling it with water when nearly full of leaves, and allowing the whole to decompose and rot.

The refuse of small gardens could be utilised by burning, and the ashes added to the soil.

The following is a list of plants that have been found to succeed in glasshouses in this country:—

Pot Plants.

Anthuriums and all Aroids.—These are easy of culture, and are not attacked by insect pests.

Artocarpus Cannonii.

Attaccia cristata.

Caladiums (pot bulbs in February, in leaf-soil, and grow near the glass).

Carludovica palmata.

C. plicata.

Dieffenbachia (and its varieties).
Fittonia argyroneura.
F. gigantea.
F. Verschaffeltii.
Ixoras.
Lomaria Gibba.
Medinella magnifica (stand the pot in water).
Nepenthes (pitcher plants).
Panax Victoriæ.
Phyllanthus nivosus.
Sanchezia nobilis (plant out in rockwork).
Sarracenias.
Many tropical ferns and cycads also grow well, but all tree-ferns are a failure.

CLIMBERS.

Some climbers, such as *Cissus discolor*, Pipers, and climbing Selaginellas can be made very ornamental by being trained round a large piece of virgin cork, placed in a suitable-sized pot to represent a tree stump, and the plant tied round. The following is a list of a few climbers suitable for cultivating in a greenhouse in this country:—

Allamanda, and its varieties	*Combretum racemosus*
	Piper nigrum
Asparagus plumosus	*P. ornatum*
A. nana	*Myrsiphyllum asparagoides*
A. tennissimus	*Pothos macrophyllia*
Akebia quinata	*P. aurea*
Bomareas	*Smilax*
Cissus argyrea	*Scubertia grandiflora*
Clerodendrons	*Stigmaphyllum citiatum*
Combretum purpurens	

ORCHIDS.

The following are a few varieties that can be grown with success in Egypt, but others may, of course, be added. With the exception of the Cypripediums and

Cymbidiums, water should be withheld after flowering until the new growth is well forward :—

Ærides, all the strong growing varieties	*Cattleya Mendellii*
Angræcum sesquipedale	*C. Eldorado*
Cymbidium Lowianum	*Cœlogyne flaceida*
C. eburneum	*Lælia autumnalis*
Cypripedium Sedeni	*L. purpurata*
C. Boxalli	*L. anceps*
C. insigne	*Oncidium Papilio*
C. concolor	*O. concolor*
Cattleya labiata	*O. flexuosum*
C. Trianiæ	*O. Cavendishianum*
C. Mossiæ	*O. trigrinum*
C. Warscewiczii	*Stanhopea insignis*
C. Sanderiana	*S. eburnea*
	Zygopetalum Gautierii

CHAPTER IX.

CACTACEOUS AND SUCCULENT PLANTS.

A USEFUL section of plants, natives chiefly of the hot dry regions of South America and the Cape. Of late years considerable attention has been paid to them, and nice collections of plants are occasionally met with.

In Egypt they are exceedingly useful, as they thrive admirably in poor sandy soil and in hot exposed places, where few other plants will grow. Many of them, such as Agave, Yucca, Beaucarnia, and Dasylirion are handsome ornamental plants, which may be improved under better conditions of soil and moisture, as high cultivation tends to develop the pulp in the leaves to the detriment of the fibre. This is preferable where plants are grown for ornamental purposes, and not commercially for their fibre. Their culture is very simple and easy, for when they are once established little or no attention need be given them.

For sunny rockeries and windy gardens they are appropriate and suitable plants; they are also useful for mixing with shrubs or as isolated specimens on lawns, while some of the handsomer species deserve a place in every garden, where the decorative nature of their foliage—and, in some instances, their large flowering spikes—are always admired.

For patches of uncultivated land that are not required for gardening purposes, clumps of Aloes, Opuntias, Yuccas, and Agaves could be planted from suckers during the rains, and, when established, would grow through the summer with little or no water, and would impart a bold tropical appearance to an otherwise barren piece of desert.

Agave. — Natives chiefly of tropical America. These plants comprise many handsome specimens, with thick, fibrous leaves; but unlike the Aloes or Yuccas they rarely form a stem. The plant usually dies after flowering, which occurs about the fifteenth year, previous to which a large number of suckers, spring from the roots, and by this means the plant is easily propagated

The Agave, which belongs to the Amaryllis family, is distinguished from the Aloe by having the stamens of its flower below the ovary, while the latter plant, which belongs to the Lily family (*Liliaceæ*), and is a native of the Cape, has its stamens above the ovary. This difference of an inferior, or a superior ovary is the distinguishing character between the *Amaryllideæ* and the *Liliaceæ*.

Amongst the representatives grown in the gardens here, are: *A. Americana, A. glauca, A. Verchaffelti, A. filifera, A. vivipara, Fourcroya gigantea*, &c. All are propagated by suckers, and some are valuable fibre plants.

Aloes—*Sabbarah.*—Succulent plants, with thick fleshy leaves, similar in form to the Agave, but without fibre. Many of the species yield a bitter juice which forms a valuable purgative, and is one of the drugs that might be cultivated commercially in this country.

Aloe vera.—This plant is grown on the graves of the Arabs, and is also hung up over the doors of their houses as an emblem of immortality. It is an erect variety, with spiney leaves of a glaucous green colour, and sometimes of a reddish-brown on the under-surface. It is not often cultivated in the gardens.

Aloe frutescens.—A nice bushy plant, suitable for pot culture or for growing in dry windy places. It is sometimes used for hedges. Other varieties of an ornamental character, such as *A. attenuata, A. indica, A. nigricans*, and *A. variegata* might also be introduced.

Cereus triangularis.—A West Indian climbing Cactus, common in the Egyptian gardens. It has large, creamy-white flowers, which open at night, and thick succulent three-cornered stems, which fasten themselves to a wall or tree by their woody roots. They are easily propagated from cuttings, and rooted portions of the stems are often used as stocks for grafting the Epiphyllum and other weak growing Cacti.

The following are a few suitable varieties worth cultivating :—*C. Bonpladi, C. hexagonus, C. grandiflorus,* and *C. tetragonus.* Care should be taken that large plants do not become over-grown or entangled, and an annual cutting-out of the crowded and weak stems should be made.

Kleinia repens.—A dwarf creeping succulent, closely allied to the *Senecio* (Groundsel), the double varieties of which form handsome garden plants. It has round finger-shaped leaves, which are covered with a milky-white bloom. A useful plant for hot sunny borders or rockeries. Propagated by cuttings in May.

Euphorbiaceæ.— A large family of about three thousand species, including many trees and shrubs, some of which have already been mentioned. The order, is widely diffused throughout the world ; but more abundantly in the hot, dry regions of South America. The succulent varieties are distinguished from the true cacti by their milky juice, which is often exceedingly poisonous. Several species yield valuable medicines, while others yield a gum from which a paint is prepared, which prevents corrosion on the bottoms of iron ships. Many varieties grow well in the gardens here.

Epiphyllum truncatum.—A common pot plant. usually grafted either on the *Cereus* or *Pereskia* stock. (The latter is a thorny climber, bearing clusters of white flowers). It can also be grafted on the *Bougainvillea* stock, which is said to impart a deep majenta colour into the flowers.

The *Epiphyllum,* which bears in the autumn numbers of dark pink blossoms, is grafted in May on stocks, which are cultivated in pots, in a compost of rich sandy soil and broken brick or sandstone, and trained on wire into an umbrella form. About the middle of June, established plants that have bloomed the previous season, should be plunged with their pots in the soil in a sunny position so as to ripen their wood, and the plants may be taken up for the verandah or house when their buds appear. Weak manure-water should be given them during their growing season.

Mesembryanthemum—Fig Marigold.—A section of dwarf creeping succulents. Natives chiefly of South Africa, to which belong some three or four hundred varieties.

M. crystallinum—Ice Plant—*Grassul.*—This species is said by some authorities to be a native of Greece, while others say it belongs to the Canary Islands.

In Egypt it grows in great abundance on the coast, and covers the low, barren hills in the neighbourhood of Alexandria, where the plants are collected into heaps in July and August, dried in the sun, and burnt by the Arabs, who use the ashes for soap-making. The plant has thick, fleshy leaves, covered with watery globules, and bears in May a number of white star-like flowers.

M. edule—The Hottentot Fig.—A species with long thick leaves, and yellow or pink flowers. The plant is suitable for covering dry banks and hot sandy places, as it thrives without any attention or water when the cuttings have once rooted, being fed from the myriads of spores which cover the leaves and supply the plants with moisture from the dew.

M. cordifolium.—A small-leaved variety, is used in the gardens for edging purposes during the summer, forming a green border from long cuttings that have their ends planted in the soil.

M. roseum, *M. nodiflorum*, *M. acinaforme*, and others are also grown.

Yucca gloriosa—Adam's Needle.—An ornamental, shrub-like plant with a large thorn at the apex of each leaf, and bearing in June a handsome spike of white, bell-shaped flowers. A very useful plant for covering banks or mounds of sand.

Y. aloifolia.—A smaller-leaved variety, often growing into a tree-like form, bearing a drooping cluster of white flowers.

Other succulents, such as *Cotyledrons*, *Crassula*, *Sanseviera*, *Opuntia*, *Ripsalis*, *Melocactus*, *Echeveria*, *Mammillaria*, and *Echinocactus* are grown, while some of the ornamental *Sedums*, which are suitable for this country, would prove useful for carpet-bedding purposes.

BULBS.

Bulbous and tuberous plants occur in large numbers in Egyptian gardens, species such as *Narcissi*, *Gladioli*, *Iris*, *Muscaria* (The Grape Hyacinth), *Allium*, *Scilla*, and many others form part of the native flora, and in the neighbourhood of Mariout immense quantities of these may be seen flowering in the spring.

Taking them in their order of merit, the Lilies ought to be mentioned first.

Lilium longifolium and *L. candidum* are, perhaps, the two best varieties for this country. They should be planted in November, in rich sandy soil, and will flower in the following April. The bulbs should never be transplanted, but allowed to remain permanently in their places, giving the beds an annual dressing of well decomposed manure in the autumn; as the longer they remain without being shifted the finer the plants will become.

L. auratum.—This species makes a bold, handsome plant in England, and one grown in peat soil in the open ground at the Royal Gardens, Kew, produced 120 flowers on a single stem. In Egypt, they should be grown in a half-shady position, but can scarcely be depended upon to produce fine flowers, and large matured bulbs only should be planted. Other species, as *L. Harrisii*, *L. speciosum*, *L. lancifolium*, *L. tigrinum*, &c., are also grown with varied success.

Gloriosa superba.—A lovely climbing Lily, lately introduced into the country from India, flowers late in the summer, and makes a pretty object when climbing over a tree-stump; also *Pancratium maritimum* (The Sea Daffodil), *Montbretias*, and *Hemerocallis fulva* (The Day Lily), with spikes of yellow flowers, which appear in May and June, and *Chlorophytum elatum*, make nice border plants. *Ranunculus*, in a sheltered spot, blooms in February and March, while the *Amaryllis*, *Vallota purpurea* and *Agapanthus unbellatus*, and *Albidus* (The blue and white African lilies), are seen at their best about the middle of April, and *Crinum amabile* and *C. asiaticum* flower throughout the summer. All the above-named species make better plants if they are not transplanted.

European Narcissi, Sparaxis, Freesias, and Ixias should be planted either in pots or in the open ground in November; the Narcissus growing, perhaps, best in the open ground, and the others in pots. After flowering, just enough water should be given to keep the foliage from withering and to ripen the bulbs. As soon as the leaves turn yellow, and part freely from the bulb, the latter should be shaken out of the soil, dried in the sun, and afterwards stored in sand until the following planting season. In some cases Narcissus will be found to throw up a quantity of leaves without flowering; but if the roots are shortened by cutting a circle about two inches from the bulb, this will often be found to check the over-growth of foliage and cause the plant to produce flowers.

Hyacinths can also be potted in November, but they usually deteriorate after the first year. When the blooms appear, an inverted flower-pot should be covered over them to draw up the flowers, otherwise the bloom will remain down amongst the leaves. Tulips and Crocus do not succeed in Egypt.

Dahlias.— The tubers of these, should be planted out in April, in rich soil, when they will grow into large plants, and continue to bloom until late in the autumn. During the growing season, manure-water should be freely given, and the plants staked securely. The tubers may be lifted in winter, but if a spring effect is only required they may be left in the ground, when they will make dwarf plants and flower in April.

Gladioli.—The bulbs should be planted about the end of December for a spring effect, and at the end of March for a summer display, the American bulbs being best for early planting. This plant, like the Freesia, Ixia, &c., can be grown from seed.

Tuberoses may be planted at the end of April, when they will flower in the late summer months; but if the bulbs are left in the ground, flowers will appear in April and May, though it is thought by some growers that the bulbs get a better rest and produce finer flowers by being annually lifted. For permanent beds the American Tuberose is the best.

Eucharis amazonica and **Imantophyllum** should be cultivated in pots for conservatory decoration. They flower best when their pots are full of roots, and should be fed with weak manure water every second or third day, as soon as their blooms appear, or a little artificial manure added. They require re-potting about once in three years, as over-potting induces the plant to make foliage instead of flowers. Plants of this kind should be kept growing in a moist shady position all the year round.

Cyclamen. — The bulbs should be planted in November, taking care that the crown of the bulb is well above the surface of the soil in the pot, and if stood in a shady, moist position, they will make nice plants and flower well in the spring.

AQUATIC AND SEMI-AQUATIC PLANTS.

These ornamental plants are unfortunately too much neglected. Many species, such as *Nymphæa* (Water Lilies), *Papyrus* (Paper-reeds), *Richardia* (Arum-lilies), *Sagittaria* (Arrow-heads), *Aponogeton distachyon* (Water Hawthorn), *Nelumbium speciosum* (The Lotus), and many other aquatics grow well in the ordinary garden fountain *(Fusghieh)*.

Papyrus antiquorum, also known as **Cyperus Papyrus,** is a native of Egypt.—This tall handsome aquatic is, apart from its beauty, interesting, because it is the plant that furnished the pith from which the paper of the ancient Egyptians was made. It is propagated from divisions, which ought to be planted in a native basket *(zambil)*, filled with Nile mud, and tied at the top to prevent the plant from floating out, and supported by large stones. The plants should be divided in February.

Nelumbium speciosum—The Sacred Lotus.—The most beautiful and graceful of all the Water Lilies, having delicate pink flowers and large round leaves, which rise considerably above the surface of the water. At the same time it is interesting on account of its remote historical associations. It was once a native of this country, and is believed to have been the "sacred beans" of Pythagoras, and the roots and seeds were eaten by the Egyptians in the time of Herodotus. Four thousand years ago it was the emblem of sanctity in Egypt and to-day it occupies the same position in the religion of the Buddhists and Hindoos. The petals of the flowers are astringent, and the fruit, which is like a large poppy capsule, cut in half, is composed of a funnel-shaped receptacle made up of a number of separate carpels, each containing a single seed, which are imported into North-West India in large quantities from China, the Malay Islands, and the warmer parts of Asia, where the seeds and stems are used as an article of food. To grow the plant the seeds should be filed, so that the *embryo* or shoot may break through the hard skin, and then enclosed in a ball of clay, and placed in the water. The seeds are, however, often lost or destroyed, as they often take a long time to germinate.

Aquatic and Semi-Aquatic Plants.

Nymphæa cærulea and **N. Lotus**—The Egyptian Water Lilies.—These plants are found growing wild in many of the water-courses of the country, particularly in those near the railway at Kafr Dowar. The bulbs should be obtained early in February, before the new leaves appear, and planted in baskets or in pots. If in the latter, care should be taken to have a large hole at the bottom and plenty of drainage, or the soil will quickly become sour. Nile mud and sand is the best compost for potting them, and the plants should be stood a foot below the surface of the water. They will soon begin to make new leaves, which will float on the surface, and the flowers, which are delightfully fragrant in the morning, are blue with yellow centres, and begin to appear in May and continuing throughout the summer.

N. pubescens, *N. rubra*, *N. stellata var. zanzibarensis*, and others might also be grown, either from seed or from bulbs. The latter is perhaps the best and surest way of raising them, and the bulbs travel well if packed in damp moss. The advantage of planting aquatics in baskets is that by the free circulation of the water the soil is less liable to become sour, and if supported between large stones the roots become firmly established before the basket rots.

Richardia æthiopica—*Calla.*—The Arum or Lily of the Nile can either be grown as an aquatic in water or in moist beds of Nile mud and manure, little or no sand being used. It is propagated by offsets of the bulbs, which grow into fine handsome plants, flowering nearly all the year round, and reaching their best about the fourth year. During the summer, an occasional watering of manure or soot water will greatly benefit the plants, and the beds should also have a dressing of manure early in the spring. The Arum Lily has a good effect when planted in water-courses or any damp places with *Calla maculata*, *Myosotis palustris* (Forget - Me - Not), and *Iris germanica*.

G

Thalia dealbata.—Native of N. America. Is also a pleasing aquatic, with erect leaves and spikes of blue flowers, which last for a considerable time. It grows freely, and is easily propagated by diversions of the roots in February. The plant is rare.

Victoria Regia.—This majestic plant, the largest of all water lilies, with leaves of enormous size, often over 6ft. across, was successfully introduced into England in 1849. It has also been grown in Cairo, and no ornamental water should be without it. The seeds should be planted in small pots in January with a little Nile mud, and placed an inch below the water, in a tank that is well exposed to the light, and heated artificially to 90deg. Fahr. As the plant develops it should be carefully placed into a larger pot. and planted out in prepared beds, two metres square, composed of Nile mud and turf, which should reach within a foot from the surface of the water, according to the size of the plant. They grow very rapidly and bloom the first season.

SEMI AQUATICS.

Alpinia nutans.—Native of India. This plant, which is known as *Zomble* by the Arabs, has leaves like the ginger, and bears during the summer terminal racemes of flowers resembling a Dendrobium Orchid. It is easily propagated by divisions, and thrives well in any damp place. The old stem should be cut down after flowering, and the plant mulched with manure.

Monstera deliciosa.—Native of Mexico. A climbing semi-aquatic, with large, perculated, spreading leaves, and flowers in the form of a deep yellow spathe, producing a long, succulent fruit of a delicious pineapple flavour, which takes a considerable time to ripen, and is known under the name of *Guista*. Propagated by cuttings.

Strelitzia reginæ—The Flamingo Plant. Native of the Cape of Good Hope. — A very handsome plant for the margin of ornamental waters, bearing large spreading leaves like a banana, to which it has a botanical affinity. It has handsome bright yellow and pink flowers. Propagated by offshoots and seed.

Ravenala madagascariensis—the Travellers' Tree. —A plant similar in form to the above-mentioned species but bolder in habit, and worthy of introduction into this country for large gardens. Other plants of a semi-aquatic nature worth growing, and already in the country may be noted; for instance, *Xanthosoma robusta*, *X. violaceum*, *Caladium esculenta*, *Hedychium Gardineanum*, *H. coronarium*, *H. chrysoleucum*, *Colocasia antiquorum* var. *Fortunesii*, *Arundo donax variegata*, *Phomium tenax* (Flat Lily), *Cannas*, &c.

CHRYSANTHEMUMS.

The number of varieties of these favourite plants that succeed in Egypt is now very considerable, and every year new ones are being introduced.

Although it is hardly to be supposed that flowers, equal in size and beauty to those we see in England can be produced here, yet with a certain amount of attention very creditable blooms may be obtained.

One of the drawbacks to the successful cultivation of the chrysanthemum is that the native gardeners have not yet learnt to grow the plants annually from cuttings, and the consequence is that old clumps flower in May much to the detriment of their autumn blooms.

Many sections such as Japanese, Incurves, Pompoms, Reflex, and a multitude of named florist varieties all succeed more or less here, according to the attention paid to them; but the number is so large and tastes differ so greatly, that in making selections, choice must be left to the individual purchaser.

CULTURE.

To obtain good plants, strong cuttings should be taken from healthy clumps in February or March, and struck in sandy soil, either singly in small pots or three or four round the sides of a 3in. pot; water them, and place the pots in a frame or a shady position until they are well rooted, and then pot on singly into larger pots. Keep the tops pinched off until they make nice bushy heads, and see that the plants are well watered and syringed to keep down the black-fly. Any buds that appear in May should be pinched off, and about the end of the month plant out in well manured beds and give plenty of water. Cut tall growing varieties to within 1ft. or 1ft. 6in. from the ground in August, and give them an occasional drenching of manure-water at their roots during the summer. The plant will then be of a dwarf character and produce good flowers in great abundance in November and December. This treatment applies only to strong-growing varieties in the open ground.

POT CULTURE.

For large flowering varieties, cuttings should be struck about the middle of March, only strong cuttings should be chosen, and not the side shoots from the base of the old stems, using light sandy soil and a little leaf mould. They should be planted singly in small pots, and placed in a shady frame until rooted. About the end of May they should be potted into larger size pots, with a little well-decomposed or patent manure added to the compost, and the pots well drained. The soil should not be filled up to the top, but an inch or two should be left so as to hold the water, and the plants placed in a half shady position and syringed morning and evening until they are well established. The tops should be pinched off occasionally until the plants become bushy, allowing from three to five stems only to remain, according to the number of flowers required. The plants will then commence to grow quickly; and by the middle or the end of June they will have filled their pots with roots, and they should then be potted into their 8in. or 10in. flowering pots, using a compost of rich sandy soil, chopped or charred turf with a few finely broken bones, and a thin layer of well-decomposed manure over the drainage, which should be carefully attended to.

When potting, the roots should in no way be disturbed, and the soil must be firmly pressed round them. The plants may then be well watered and plunged with the soil up to the rim of the pot in sunny beds. They should be irrigated according to their requirements, which will be about every second day, and alternate waterings with manure and soot water, should be given every four days. Too much water will be found to turn the foliage yellow.

Each shoot should be tied out separately early in September with a neat stick, and the plants syringed morning and evening so as to prevent the blight and black-fly from injuring the young shoots.

As soon as the buds appear, the side ones ought to be removed, leaving the terminal ones only three, six, or eight flowers being sufficient for a specimen pot plant. The plants, as they come into bloom, should then be taken up, their pots washed, and stood in suitable positions in the house, verandah, or the garden.

ANNUALS.

Since seeds can be obtained with the illustration of their flowers on each packet, the amateur has no longer any difficulty in making a suitable collection, and arranging them according to their height and colour.

Many varieties, such as Phlox, Gaillardias, Poppies, &c., will sow themselves, and come up year after year, but most of the species mature their seed so rapidly that the old seed is not worth keeping for the following year, and fresh supplies had better be annually obtained from a reliable firm so as to ensure success.

The seed should be sown in boxes or pans from October to January, the display in Cairo being considerably earlier than that of Alexandria, though the light soil of the latter place suits them better. Sow thinly in sandy soil, give plenty of drainage, and place the seed-pans in a sheltered spot. Pot the seedlings off, or plant out as soon as they are large enough to handle, care being taken not to injure the roots, and to have a little soil attached to them, and plenty of water given. The late afternoon is preferable to the morning for placing out the young plants, as they have time to recover before the sun reaches them, but if a showery day can be chosen, so much the better.

Antirrhinums (Dragons' Mouths) should be sown in August or September if flowers are desired in the early spring, while a later sowing will produce plants that will flower through the summer.

Petunias, *Nicotiana affinis* (The Tobacco), *Zinnia elegans*, *Linaria*, Stocks, *Salvias*, Perennial Sunflower, and *Canna indica* are useful for beds for summer flowering. *Eschscholtzia californica*, *Tegates erecta*, *Tropæolum majus*, Lobelia, *Linum grandiflorum*, Larkspur, Mignonette, *Nigella*, *Silene*, *Portulaca grandiflora*, Shirley Poppies, Verbenas, *Helichrysum*, Calliopsis, Scabiosa, and many others may be sown in patches in the open ground for spring and early summer.

Some varieties, such as Gaillardia (which are useful for cutting during the winter) Calliopsis, Scabiosa, &c., become perennials, but the large majority gradually die off as the summer sun gains power.

Sweet Peas may be sown thinly in pots and afterwards planted out if required for border decoration.

Ornamental grasses are sometimes difficult to grow by themselves if sown too thickly, but they will often grow well if mixed and sown broadcast with Shirley poppies or any light-growing annual.

BALCONY GARDENING.

In town houses, where plants are cultivated on balconies, they will be found to thrive, and flower much better if grown in boxes rather than in pots, for by the latter method the young roots are destroyed by being scorched by the sun as soon as they reach the sides of the pot, and unhealthy plants are consequently seen. If, however, boxes are used, their roots can grow unchecked, and a clean and tidy balcony full of healthy plants, instead of one filled with sickly specimens in pots of all sizes, may be maintained.

In arranging a balcony for growing plants, boxes of a suitable size should be made to fit into convenient places. These should be tin-lined or tarred inside to prevent the wood rotting. Large holes being made at the bottom, and the boxes raised a few inches on blocks to allow the water to pass away freely.

The outside may be covered with the bark of the Date Palm or virgin cork nailed on to the wood and varnished, or a coat of green paint could be given. Large pieces of concave pottery should be placed over the holes inside the boxes, and smaller pieces, to the depth of two or three inches, should be laid on the top for drainage covering them with a thin layer of dried grass or long manure to prevent the soil from washing down, and then fill the boxes to within two inches of the top with a compost of rich, sandy soil and a little decomposed manure, which can be obtained from a garden or nursery. The boxes should be well watered and filled with hardy, dwarf plants or annuals, with a plant of Stephanotis, Hoya, or Ivy trained round the rails or on bamboo canes up the wall; and trailing plants, such as ivy-leaf Geranium, *Saxifraga sarmentosa, Tradescantia discolor*, &c., may be allowed to hang over the side. The boxes should be watered daily, and the plants syringed morning and evening when the sun is off them and a dressing of fresh soil and manure should be added when the compost in the boxes becomes exhausted, and the plants sheltered in winter by a mat.

PLANTS SUITABLE FOR A SHADY OR MOIST POSITION.

Adiantum Capillus-veneris (on rocks)
Æchmea fulgens
Agapanthus umbellatus
A. albidus
Anemones
Amaryllis
Arum Lilies
Aspidistra lurida
Caladium esculentum
Chlorophytum elatum
Cyclamen
Hemerocallis flava
H. fulva (Day Lilies)

Hoya carnosa (Wax Plant)
Hydrangea
Ivy
Liliums
Monstera deliciosa (French, Guista)
Narcissi
Nephrolepis exaltata
N. acuta
Pancratium maritima
Peperomia clusiæfolia
P. marmorata
Pteris longifolia
Phormium tenax
P. tenax variegata
Polypodium aureum
Reinwardtia trigynum
Ranunculus
Salvia coccinea
Saxifraga sarmentosa (Mother of Thousands)
Strelitzia reginæ (Flamingo Plant)
Thalia dealbata (in water)
Tradescantia (Spiderwort)
Vinca rosea
V. alba
Xanthosoma violaceum

LIST OF PLANTS SUITABLE FOR A WINDY POSITION.

Cacti and Succulent plants, including
Agave
Aloe
Cereus
Cotyledon
Crassula
Echinocactus
Euphorbias
Gasteria
Mammillaria
Mesembryanthemum
Opuntias
Phyllocactus
Rhipsalis
Stapelia, &c.

Arundo donax
A. donax variegata
Beaucarnea recurvata
Buddleia Madagascariensis
Centaurea candidissima
Cineraria maritima
Cissus rotundifolia
Dasylirion acrotrichum
Eleagnus hortensis
Gazania splendens
Kleinia repens (*Senecio*)
Lonicera japonica (Honeysuckle)
Lycium europæus (Sea Buckthorn)
Nicotiana rustica (Persian Tobacco)
Oleandra
Œnothera (Evening Primrose)
Pelargonium, Ivy-leaf and zonal varieties
Pittosporum undulatum
Ricinus communis (Castor oil)
Tamerix arborea
Yuccas

CHAPTER X.

FRUIT AND VEGETABLES.

FRUIT.

THE fruits cultivated in Egypt are varied and plentiful: Oranges, Mandarins, Custard Apples, Figs, Grapes, Melons, Apricots, Bananas, Pomegranates, and others give abundance of fruit during their season; but the style of their cultivation is somewhat primitive, and very little is done to improve their quality or introduce new varieties. At the present time, fruit-growing holds an important place in the industries of the world, and if the matter was taken carefully in hand much might be done to improve the standard of fruit-growing here by grafting strong native stocks with new varieties, systematic pruning and manuring, and also by the introduction of many tropical fruits that are comparatively unknown in the country, a list of which will be found at the end of this chapter.

Soil, and the selection of a suitable site, is of the chief importance, and some shelter, either by a wall or hedge, should be afforded for the protection of early blossoms from the strong sea winds for places on the coast; and those inland from the cold winds from the desert.

The alluvial Nile soil is of course preferable to one of a poor sandy nature, but if shelter can be obtained, good results may be expected in the latter if, at the time of planting, the holes are dug and filled with good rich soil, and sufficient room is allowed for each tree to develop; while exposed places, where little water can be obtained, could be devoted to Grapes and Figs, the latter being planted in a trench and the soil thrown up into a bank on the exposed side, so as to form some shelter for the vines.

Anona squamosa—Custard Apple—*Guista*—Native of the West Indies. A well known delicious fruit, commonly cultivated in Egyptian gardens. The trees, which are often of a dwarf shrub-like form, grow readily from seed. It is the latest plant in the country for transplanting, which should be done about the end of March or early in April, when the plant looses its leaves. Large trees transplanted with balls of earth at their roots will fruit the same year. The trees should be well manured in the winter, and the ground lightly cut up. Water once a week in dry weather until the buds appear, when water should be withheld, otherwise the flowers will drop off, and again occasionally when the fruit has formed. At the time of ripening, the trees should be looked over every morning and the fruit picked as it becomes soft, or it will fall and be destroyed. The trees should never be pruned, but occasionally a branch may require cutting back or thinning out. A variety known as *Balady* (or native), though inferior to the former, is also grown.

Carica Papaya—Papaw.—A native of tropical South America. In this country it is grown as an ornamental plant. but I am not aware that any use is made of the fruit. The plant grows in the form of an Aralia, having a single stem rising eight or ten feet high, with a bunch of leaves at the top and a cluster of green, pear-like fruit hanging at their base. These should be thinned out soon after they form, in order that those that remain may come to perfection, and the flowers that continue to develop should be picked off. The trees should be frequently watered, and the fruit, which ripens in the autumn, is agreeable and wholesome when eaten with sugar, or preserved.

Plants can be propagated from seed sown in the spring, but as they are of two sexes they will not all produce fruit. The tree is said to have the property of rendering tough meat tender by washing it in water impregnated with the milky juice, or by wrapping it in the leaves, which causes a separation of the muscular fibres.

Citrus aurantium—Orange-*Bortugan*.—There are three varieties of Oranges grown in Egypt. The *Masri* or Cairo Orange, a juicy, thin-skinned variety; the *Dhum* or Blood Orange; and the *Sharmi* or Jaffa Orange, a large, thick-skinned variety. A sheltered position, where the blossoms can fully develop, is one of the chief things in Orange cultivation. Very fair crops may be obtained on the poor soil of Alexandria and Ramleh, though constant attention must be paid to manuring, which should be well decomposed and placed in a trench dug round the trees.

In the rich alluvial soil of Cairo the trees grow to a larger size than those near the coast, and produce immense crops of fruit. If, however, the soil is too heavy, lime rubbish should be mixed with it. Oranges may he grafted in February and also at the rise of the Nile on stocks of either the *Naring (C. Bigaradia)*, The Citron *(C. media)*, or the Lemon *(C. Limonum)*. The first is largely used by the native growers, from whom grafted trees can be purchased at a very reasonable price. Blood Oranges are said to be grafted on two-year-old stocks of the Black Mulberry.

C. Limonum— The Lemon — *Limoun Balady*. — The common Lemon fruits three times a year, and it is not unusual to see ripe and half-developed fruit, together with the flowers on the tree, at the same time. The culture is similar to that of the orange, and the trees should be occasionally thinned and strong leading shoots shortened and suckers cut off.

Two other varieties, namely, *C. decuma*, Small Lemon *(Limoun Hindy)*, and *C. Limonum* var. *dulcis Risso (Limoun Helon)*, or Sweet Lemon, are also grown. All may be raised from seed or propagated by layers.

C. nobilis—Mandarin—*Yousef Effendi.*—In the more modern gardens the Mandarin is sometimes grown with round bushy heads similar to an ornamental tree; and, like the orange, it requires a sheltered position and a prepared soil if good crops are to be obtained. Trees of this kind should have their tall shoots pruned back in February, and care should be taken that the soil is not too wet or too dry when the fruit is just forming, otherwise much of it will fall. Where there are very heavy crops a considerable quantity of fruit will be sure to drop during the hot weather, as the tree will be overladen, and nature will intervene in order that it may not become impoverished. It might be, perhaps, advisable, under such circumstances, in small gardens, to go over the trees and thin out the crowded fruit.

Irrigation should be done every fourteen days during the summer, the time for which can be ascertained by cutting up a piece of the soil to see if it is dry beneath. Manure-water occasionally given at this time when the soil is not too dry will also assist the trees.

In large plantations where little pruning is done, all dead and useless wood should be cut out in April, and a dressing of cow manure may be given every second year in February to induce the roots to come to the surface.

C. media—The Citron—*Touroung Balady.*—Commonly grown in the gardens as an ornamental tree. It has bold, dark green foliage, and its large pendant fruits of the size of an ostrich's egg are very ornamental. It is easily propagated by ringing or layering.

All the *Citrus* tribe and plants with aromatic leaves, such as the Sweet Bay, Croton, &c., are attacked during the spring and summer, especially in gardens near the sea, by a small brown scale. This I have succeeded in destroying by syringing with a solution of a wine-glassful of petroleum and a small piece of soft soap to a pail of water, which should be kept stirred while using, and apply in evening with the syringe. Though this remedy is effectual in killing the scale, they still remain tightly glued to the leaves.

Eriobotrya japonica—Loquat—*Beshmeyleh.*—Native of China and Japan. A small tree with large rough leaves, bearing clusters of plum-like fruit of a yellow colour, and containing a large brown seed. The trees are sometimes attacked by a maggot, which eats away the pith in the stems. When this occurs it will be found beneficial to paint the stems with a mixture of equal parts of whiting, petroleum, and soot, and to cut up the soil in the early spring.

The Loquat rarely makes any superabundant growth, and therefore does not require pruning. A good dressing of manure ought to be given in January, and water frequently throughout the summer. It is propagated by seed, which should be sown soon after it is ripe, as it quickly loses its germinating power.

Fragaria vesca—Strawberry—*Frowley.*—Strawberry-growing in the gardens on the coast has, during the last few years, greatly increased, and considerable areas are now under cultivation for the markets. The larger varieties are unfortunately unnamed, but this is possibly done to prevent a glut in the market, and growers even go so far as to destroy all surplus runners, so as to prevent the variety becoming common. Many experiments may therefore be made before a variety suitable for the country is obtained, but, as a rule, those that produce the least foliage invariably give the best crops on light soil, while the Black Prince is thought to succeed best on a heavier soil.

The beds should be made in August, the surface soil being well dug, without turning the sub-soil. Well-decomposed horse manure or night soil and ashes should be applied heavily, and the beds made three metres wide in order to facilitate irrigation.

In preparing the runners, those nearest the parent plant and the largest of the remainder should be taken and planted about 9in., with 1ft. 3in. between the rows. After planting, water ought to be given every day until the young plants are established, and later on once every six or seven days, according to the weather and requirements.

About the end of February, or early in March, is usually the time for the first fruit to ripen ; though, in mild winters, it is not uncommon, even before Christmas, to see Strawberries, and they continue to give fruit until the end of June. During the fruiting season, the plants should be watered after every picking, as water must never touch the ripe fruit, which should be supported on pieces of tiles to keep them clean, and in the neighbourhood of Cairo it is necessary to net over the fruit to protect them from the crows.

In the middle of June, when the runners appear, a light framework should be erected about four feet from the plants, and covered thinly with palm-leaves. This will induce the runners to grow more freely.

New beds should be made every second year, as the crop the first year is by far the best, and where sufficient runners can be obtained it is advisable to make new beds annually. The most favourable position for planting is under a south wall, as the early blossom is then protected from the heavy rains and north-west winds.

The Alpine Strawberry—*Frowley Balady*—is also largely grown, and can be obtained either from seed or runners. It grows freely in light sandy soil, and produces an abundant crop of small well-flavoured fruit in April and May. A space should be devoted to this variety in every garden.

Ficus Carica—Fig—*Teen.*—The Fig is cultivated in large quantities in the neighbourhood of Alexandria and Ramleh, and bears abundant crops from June to September. Those in Upper Egypt ripen a month earlier, about the end of May.

The trees delight in a rich, sandy soil, and are usually irrigated in the spring when the foliage appears, and again at the rise of the Nile in August. Frequent watering causes the fruit to drop. Their culture is very easy, and with the exception of cutting out weakly shoots, or dead wood, no pruning is required. Some cultivators recommend a slight dressing of manure every second year, which should be applied in February and lightly dug into the soil.

The Fig can be propagated by cuttings of ripe wood, half a metre long, which should be taken in the early spring and planted in sandy soil in a slanting direction, with their heads just above the ground.

The following are the varieties, under their native names, that are commonly grown:—

Saltarne—The Sidi Gabr Fig.—A long, large, red variety used for drying and preserving.

Cou-met-re—The Pear Fig.—This variety is similar to the former, but whitish instead of red. It is considered to be the sweetest of all.

Hub-bas-hee—A small black round variety.

Balady.—Round and flat.

Utsee—Round, flat, and red.

A-gay-eé.—A very sweet, green, flat, and round variety. Others of a good quality might also be introduced.

Mangifera indica—The Mango.—A somewhat dwarf tree with oblong dark green leaves like a Ficus, and bears in April and May erect terminal panicles of small whitish flowers. A few nice specimens exist in the gardens of Cairo, but the fruit is of little value. Much, however, might be done by grafting and cuttings, as there are nearly as many varieties of the Mango as there are of apples, and consequently a considerable difference in the quality of the fruit.

Musa sapientum—Banana—*Mouz.*—This dwarf variety is largely grown in Lower Egypt in plantations, and also in private gardens. It requires a rich, moist soil and plenty of manure. The fruit, which is borne at all seasons of the year, should be cut just before it ripens, and hung in a dark room, where it will become mellow and well flavoured. After fruiting, the old stems should be cut away, the weakest of the suckers thinned out, and a good surface dressing of cow-manure given. The Banana is propagated by suckers which are usually planted from two to three metres apart in rows, and in well-manured ground, each plant being allowed to develop three fruiting-stems. A thinning of dead leaves should be made at least twice a year.

M. rosacea.—Native of India. A tall, handsome species, with red stems and large, graceful leaves. It is usually grown more as an ornament than for fruit. It bears a small, thick, dark-skinned fruit of a rich flavour, but does not produce so good a crop as the former variety.

M. Ensete.—A West African species of an exceedingly ornamental appearance, grown chiefly as a specimen plant, on lawns. It has a thick stem, and large, handsome leaves, and requires two or three years or even longer to flower. It may be propagated from seed, but the fruit is useless, and the plant should not be exposed to heavy winds, or the beauty of the foliage will be destroyed.

Opuntia Ficus indica—Indian Fig—*Teen Shok.*—The fruit, which is considered very wholesome and nutritious, is borne in great quantities on large cactaceous shrubs. It ripens in the middle of July, and should be gathered early in the morning and eaten at once, as it soon turns soft after picking. Care should be taken when peeling the fruit that the minute thorns do not enter the flesh, as they are difficult to extract. The plants, which make an excellent hedge for a dry windy position, can be grown from cuttings taken in May, and no attention need be given them after they have rooted. The leaves when boiled can be used instead of size for whitewashing.

O. maxima and other varieties are also grown for decorative purposes.

Physalis peruviana—Cape Gooseberry—*Halwah.*—The plant which belongs to the Potato family

(*Solanaceæ*) is a dwarf herbaceous perennial, naturalised in the neighbourhood of Alexandria and Damietta. The fruit, which is the size and colour of a cherry, is developed in a leafy appendage. It can be eaten raw, or made into a preserve, and is easily propagated by seed.

Prunus armeniaca — Apricot — *Mich-mich.* — Like the fig, this fruit is largely cultivated throughout Egypt. The trees, which bloom in February, delight in a rich porous soil, and a sheltered position. The common variety is grown from seed, and fruits about the fifth year, they should never be pruned; but only the dead wood removed. In cutting away large suckers it is advisable to bind over the wounds to protect them from the sun until they have healed, or otherwise they are likely to canker. As they are surface-rooting trees, the ground beneath should not be dug; but a layer of decomposed manure, and garden refuse spread over the surface and well watered will greatly benefit them. At the time of flowering a sweet smelling herb known as "Sheer" (*Gypsophila*) is tied in pieces of palm fibre and hung in the branches to prevent the flowers from being injured by insects.

Four varieties are grown in the country, namely :—

Balady, the common variety, of a poor quality, bears a heavy crop, and is the earliest of all. The kernels are roasted and pounded by the Arabs, who take them in the early morning with water as an emetic. The fruit is ripe in May.

Cal-la-by, a small variety fruiting a month later.

Hum-oi-de, the largest and best variety, coming into season about the middle of June. It should be grafted on the three-years-old Balady stock, and can be eaten raw or made into preserves.

Mich-mich Sharmi, a Syrian variety, with large, long fruit. Is imported in July.

P. persica — Peach — *Khokh.* — A few common varieties are occasionally grown, but are usually much neglected. As a rule the European varieties usually degenerate, but many of these might be made to succeed if grafted on strong plants of the former stock or the Almond *(Amygdalus communis)* and other established varieties, and thus the quality of the Peach, Plum, Nectarine, Apricot, and other sections of this tribe in Egypt might be considerably improved. All healthy trees should have the branches that have fruited pruned back, and water should be withheld when the trees are in flower and until the fruit attains the size of a pea, water then once a month will be sufficient. Australian varieties would no doubt succeed in Egypt.

P. domestica—The Plum.—Two varieties of plums are grown in the country, the above-mentioned, known under the name of *Hum-oi-de*, and *P. divaricata*, known as *Balady*. The former variety is considered the best, and should be grafted on to the latter; little attention is, however, paid to their cultivation, and as the fruit is consequently poor they are not often grown. Good varieties might be introduced from Japan.

Punica granatum — Pomegranate — *Roumman*. — Often grown in the gardens on account of the decorative nature of its flowers. The plants can be raised from seed sown in October, and by cuttings or layers, which should be made in February. Rich, well-manured soil suits them best, but they require less water than other fruits. Young wood and suckers which often grow at the foot of the trees should be cut away, and the branches that have fruited should be pruned back in the winter when the leaves fall.

The following varieties are grown :—

Ma-le-ce, a large white variety.

Balady, a blood-red variety, used for preserves.

Ma-grub-be (Tunisian), red and white.

Sharmi (Syrian), the fruit of which is pink.

The bark of the tree is used for tanning morocco leather, and the roots are said to be an excellent vermifuge.

Pyrus communis—Pear—*Coumetre*.—Fruits of this kind, including the *P. malus* (Apple), *Tofar*, are occasionally met with in the modern gardens, and although a variety may sometimes be found that will bear fruit for a number of years, yet their culture can scarcely be called a success, and they either deteriorate by blooming at all seasons of the year, or are destroyed by a maggot which attacks the Loquat and eats away the pith in the branches.

Vitis vinifera — Grape — *Aneb.* — Some fifteen to twenty varieties known under their native names are said to be grown in the country. Most of them are raised from cuttings, but few, from want of attention, can be said to be a complete success, and the immense influx that is annually poured into the markets during the summer from neighbouring countries, may, perhaps, account for steps not being taken to improve the local varieties. Like the Fig, various opinions are expressed as to the proper system of cultivation, and the quantity of water the vines should receive.

The best grapes grown in Egypt are those that are irrigated but twice during the year. Once after pruning in February just as they begin to break, and again after the fruit has formed, as frequent watering causes the grapes to shank and drop, and promotes an overgrowth of foliage.

Vines may be grown from cuttings about half a metre long, taken in February, and planted in the ground in a slanting direction so as to enable the cuttings to root up the stem, leaving two eyes above the surface. These should be trained over lattice work, and pruned closely back to two eyes in February. Old plants may also be pruned in the same month, and a good dressing of decomposed cow manure and night soil should be lightly forked in about the roots, and afterwards well irrigated. They should be pruned closely into the main stems, leaving two eyes only. Some growers replace the old rods every third year.

For ground plantations the cuttings ought to be planted three metres apart in the lines, and the same

H 2

distance should be left from row to row. They should be pruned back to three eyes the first year, and the annual pruning should be done closely, so that the vines represent a bush.

The following are some of the chief varieties grown in the country:

Showish. — This is the variety chiefly grown in Egypt, and is the best for ground vineries. When grown on the dry system, as seen at Gabarri, the fruit is a large, long grape of a dark-red colour; but when badly grown, which often arises from too much water, as previously mentioned, it usually becomes a dullish green, and the bunches are long and thin.

Frowley or Strawberry Grape. — A variety with slightly-lobed palmate leaves, which are white on the under surface. The fruit is round and of a deep red colour, with a fleshy, strawberry-like flavour. Its origin is said to be Greek.

It is best when grafted on two-years-old plants of the Showish variety, which should be done early in February; but when grafted in the *cleft* form on the stems of old vines they will ripen their fruit earlier than if grafted on young plants.

Bou-lis-se.—The earliest of all. It has small, round, black fruit, borne in a compact mass on the bunches. It is said by some to have a salty flavour.

Turkey.—A large, round, white variety.

Ben-ar-te.—Red, slightly rounded, and stoneless.

Muscat.—Red, long, and large.

Gaz-as-se.—Red and long.

Noi-wy-me.—Round and white.

Dron-khun.—Red and round.

Many of the best varieties from England might also be introduced.

A list of new fruits likely to prove successful in Egypt:

Anona muricata—Sour Sop.
A. cherimolia—The Cherimoyer.
Arduina bispinosa—Natal Plum.
Averrhoa carambola—Chinese Black Currant.
Blighia sapida—The Akee.

Chrysophyllum cainitum—Star Apple.
C. oliviforme—The Damson Star Apple.
Dillenia speciosa—Indian Green Apple.
Diospyros Kaki—Date Plum.
Eugenia Jambosa—Malay Apple.
Falcourtia Ramontchi—Governor's Plum.
Mammea americana—Mammee Apple.
Nephelium Lichi—The Lichee.
Persea gratissima—Avacado Pear.
Tamarindas indica—Tamarind.

VEGETABLES.

It is generally admitted that the ordinary vegetables are cheaper to buy than to grow, but some of the better kinds might well be cultivated if there is sufficient space in the garden.

Agaricum campestris — Mushrooms. — Several attempts have from time to time been made to cultivate the mushroom in Egypt ; but owing partly to the want of knowledge and a suitable place to grow them in, and partly from the difficulty in obtaining the spawn, the efforts have generally proved unsuccessful, although, if their culture could be carried out with success, the undertaking might be looked upon as a profitable one.

The following directions for their culture are given by Mr. P. W. Gibson, late of the Garden Department, Hyde Park, London, who has had several years of horticultural experience in Egypt :

To make the beds, fresh horse-droppings should be collected and placed in a heap to ferment ; after a few days these should be turned over and a little water added if dry, and in a day or two they should be taken into a dark shed and made into beds, 2ft. wide and 1ft. thick, the top of the beds being slightly smaller than the bottom. These beds should be looked to occasionally until the temperature remains stationary at 77deg. Fahr., when the spawn, in pieces the size of hens' eggs, should be inserted just below the surface. As soon as the thread-like filaments appear, an inch of light sandy soil should be spread over the beds, which should be previously moistened by a slight sprinkling of water if dry.

In eight or ten weeks the mushrooms will appear, when they should be cut off carefully so as not to injure the roots. The crop will continue for some months.

Asparagus officinalis—Asparagus.—This plant is found growing wild in the Nile valley. In the gardens its cultivation is extending very rapidly, but generally the crops are poor and the heads are very thin and small in size.

The beds should be prepared about Christmas, in the following way: First, the soil ought to be dug about 3ft. deep, and 1ft. of clinkers, pottery, or broken bricks thrown in, in order to facilitate drainage; then a foot of the ordinary soil placed over the drainage; and, lastly, the remaining space filled with rich sandy soil, which has previously been well mixed with old decomposed manure.

The beds are made two or three metres wide, and should not be raised as in Europe, but kept low in order to facilitate watering by irrigation. The seeds should be sown early in March in small beds of their own, or can be planted somewhat earlier in pots. If the former system is preferred, a few pots of seedlings should always be kept to fill up vacancies. When the seedlings are a year old they may be planted out in their permanent beds. They should be placed six inches apart, and a little salt might with advantage be added to the soil. Water must be given every second day, and the beds never allowed to get very dry. The plants should not be cut the first year after planting out, but some people recommend picking off the seeds without injuring the foliage, in order to strengthen the crowns.

The second year a crop may be expected, and cutting should continue from March until July, though the finest crop is in April and May. To obtain white well-bleached heads, the Arab growers recommend pieces of hollow bamboo cane being placed over the crowns to protect them from the sun. In January the beds may be lightly forked up on the surface, and a good coating of manure added. Some recommend sea water or salt to be given to the beds in February before the heads appear, while others say an occasional drenching of manure water in March is very beneficial.

Brassica rapa—Turnip.—The seed should be sown broadcast during the rains, for unless they grow quickly they become stringy and hot. Carrots and Radishes may also be sown in the same way.

Cynara scolymus — Artichoke — *Kharchouf.* — The plants require a rich sandy soil; seaweed or saltpetre being recommended as a manure for them. They may be raised from seed sown in July, or divisions in August.

Phaseolus americana—American Bean.—This Bean, which bears an excellent crop, is considered best for a summer supply. They should be planted early in April, in trenches a foot apart in double rows, leaving 4ft. or more between each trench, as they grow to a considerable height. About the middle of June the crop is ready, and where quantities are grown pickings may be made daily until the end of November. The crop requires to be heavily manured.

P. vulgaris—French Beans.—The seed should be planted a foot apart on ridges, about the middle of April, and water given every four or five days. The crop may be picked in about five weeks.

Butter Beans.—This variety may be planted at any time from April until August in trenches, each seed being placed a foot apart, and four or five feet left between the rows. They should have strong sticks placed to support them about a month after planting, and will be ready for picking in eight or ten weeks.

Pisum sativum—Peas—*Vercilla.*—The ground for peas should be dug deeply, but not manured; a little saltpetre added to the soil is said to give good results, but the success of the crop depends much upon the rainfall. The soil should be thrown up in ridges, on which the peas may be planted three or four together at intervals of about six or eight inches, and supported with the canes of *Arundo donax* (known as "Bousa") or dried branches as soon as the tendrils appear. The first sowing may be made at the end of September, and continue, at intervals of one month, until the beginning of December. The pods should be cut off with a pair of scissors, and not wrenched off with the hand, while the best should be saved for the next season's sowing. "Early Longpods" and "Marrowfats" are two good varieties. Some growers recommend a change of seed every three years.

Solanum tuberosum—Potato.—The first crop may be planted the last week in September. The seed should be soft and pliable to the touch, and the eyes just started. When planting, place the smallest end, which contains the strongest eye, upwards. They should *not* be cut, or they will turn black and the eyes will not break. The plants should be ten inches apart, and three feet from row to row. Water should not be given until a fortnight after planting, and when the plants are four inches above the ground, they may be earthed up and water given once a week in dry weather.

Planting may be continued up to the end of January as the ground becomes vacant, and a light dressing of road sweepings may be given when preparing the soil.

In three or three-and-a-half months after planting the crop will be ready for lifting. Varieties such as the French Round or red kidney potatoes, and the "Early Shores" (a rough-skinned variety) are, perhaps, the best.

S. melongena — Egg Plant — "Beydingan."—The plants are raised from seed, sown in well-manured ground in April.

The time for sowing other vegetables will be found in the "Garden Calendar."

THE GARDEN CALENDAR.

January.—Borders should be dug up and the soil left open. Uneven growth on trees and shrubs should be cut back.

Path-making, rockeries, raising of mounds, and work where soil has to be carried, may now be pushed forward.

Paths should be rolled after each rain and lawns cut and rolled when dry. Chrysanthemums may now be cut down, divided, and planted out in nursery beds, or stored away in their pots for cuttings in April.

Give a surface dressing to Loquats, Mishmish, Figs, Peaches, Plums, Mandarins, and Oranges.

Prune the first batch of La France Roses that are sheltered about the middle of the month.

Keep Maidenhair Ferns and Anthuriums drier than usual, and place in a light position Orchids that are throwing up new flower spikes.

Climbing plants in houses that are dormant may now be pruned and washed. The paint and inside glass in the greenhouse should also be cleaned with warm, soapy water. Close the houses early in the afternoon. The last crop of Peas may now be sown and late Potatoes planted.

February.—Transplant Date and Ornamental Palms about the 8th of this month. Prepare ground for permanent lawns of Neguil. Fruit and ornamental trees, shrubs, and climbing plants may now be removed.

Prune Roses, Mandarins, Vines, and Mulberries, about the middle of the month. Give rose beds, vine borders, and bulbs a dressing of manure. Plant out Pansies, Asters, Stocks, &c., that are large enough.

Large branches of Tamerisk and Lebbek may now be planted for forming new trees.

Roses may be budded towards the end of the month; Vines, Oranges, and Peaches may also be grafted.

Sow seeds of Poinciana, Eucalyptus, Acacia, Pinus, Cassia, Tecoma, and pot the first batch of Caladium bulbs. Maidenhair Ferns may now be re-potted, and

Azalias, Narcissus, Cyclamen, as they come into bloom, should be removed into a cool greenhouse.

Water-lily bulbs may now be obtained and potted.

Sow Spinach, Lettuce, and Endive.

March.—Amaryllis, Imantophyllums, and other bulbs that are coming into bloom, should have manure-water given them, and Hyacinths covered with inverted pots.

Lawns and paths should be frequently rolled, hedges clipped, and dead wood cut out of trees. Filfil trees should also be cut back. Syringe the foliage of Crotons in houses and leave a ventilator slightly open at night.

Roses that have not been budded should now be finished.

Caladiums that have well started may now be potted on, and fresh moss added to the surface of Anthuriums.

Slugs and snails should be caught by putting pieces of cabbage leaf and potato on the stages in the greenhouse, which should be examined early each morning, and orchid spikes protected by pieces of cotton-wool. Cuttings of Heliotrope, Pittosporum, Jasminum, and all bedding plants may be made. Gladioli may now be planted.

April.—Roses will now be at their best. Pick off rose-beetles and dead flowers. Give climbers in the greenhouse a top-dressing of manure, and manure-water to Cycads. Syringe newly-potted palms.

Buddleya Madagascariensis and *Lantania hybrida* should now be cut back and Chrysanthemum cuttings taken. Give newly made lawns a light coating of manure, and keep the turf well cut and rolled.

Towards the end of the month the garden will require watering twice daily with the hose. Plant Tuberoses; Bamboo roots may also be planted, and Poinsettias and *Montanoa grandiflora* should now be pruned. Sowings of French Beans, Vegetable Marrows, Tomatoes, and Egg Plants may now be made.

May.—Summer bedding plants may be planted out, and Libea planted for summer lawns. Neguil may also be sown for permanent lawns, and cuttings of Mesembryanthemum and all cactaceous plants should now be taken.

Water copiously newly-planted palm trees, shrubs, &c. Those that have lost their leaves should be occasionally syringed.

Dahlias for summer flowering may be planted.

Trees should be propagated by ringing, &c.

Glasshouses should be well ventilated and supplied with plenty of moisture. Shading also will now be required.

Sowings of American, French, and Butter Beans, Vegetable Marrows, and Egg Plants may also be made.

June.—Chrysanthemums that are forward should be shifted into their flowering pots.

Flowering Cacti in pots should be plunged in sunny beds, and all plants in the greenhouse, with the exception of Ferns, Anthuriums, Orchids, Caladiums, and very tender plants should be stood outside under a light shading of palm leaves; and those that have filled their pots with roots should be potted on.

Strict attention must now be paid to watering, and plants syringed night and morning.

Continue to sow American, French, and Butter Beans, Egg Plants, and Vegetable Marrows.

The ventilators in the houses should now be left half open at night during the next four months.

July.—Lawns from grass seed that became brown and dry at the end of last month should now be cut up and allowed to remain until September.

Large trees should have their summer pruning, also pruning the large roots, and mulching with night soil may now be done. Trenches should be made round all large specimens for the irrigation and manure water. Sow Artichokes, Cabbage, and the last sowing of Beans.

August.—Pot Crotons, Chrysanthemums in the open ground should now be cut back. Supply Dahlias, Hollyhocks, and all flowering shrubs with manure-water.

Turf and manure should be stacked for spring potting.

Carnation and Croton cuttings may be taken.

Strawberry beds should now be made, and any grafting or budding should be done at the end of this month. Hedges may be clipped. Rose-beds mulched with manure and the water supply increased. Sow Cabbage, French Beans, and Cauliflowers.

September.—Budding and grafting should be finished this month. Lawns should be prepared for grass seed.

Chrysanthemums in pots should have their shoots tied out. Plants in pots should be looked over and cleaned, and continue to give them manure-water twice a week.

Plant first batch of Potatoes the last week this month. Sow Broad Beans, Spinach, Lettuce, Beet, and Endive.

October.—Annuals may be sown about the middle of the month. Grass lawns from seed should be cut and rolled. Plants in pots may be cleaned and got ready for housing next month. Grass lawns from seed should be sown and finished this month, and hedges and borders clipped.

Climbing plants may be gone over, and untidy shoots cut back or layered. Sow Turnips, Broad Beans, Spinach, Endive, Lettuce, Carrot, and Beet.

November.—Plants that have stood out during the summer should be placed in the glasshouse this month.

Lilies, Cyclamen, Narcissus, and all spring flowering bulbs may now be potted. Continue to sow annuals for a late show. Manure-water should be discontinued, and the houses closed just before the sun is off them, and heavy shading removed.

Datura and summer flowering shrubs may be pruned back. Paths should be repaired, and the drainage attended to. Sow Celery, Beet, and Peas. Continue to plant Potatoes.

December.—Paths and lawns should be swept and rolled, and borders raked, and things kept neat and tidy.

Large Fan Palms in windy gardens should have their leaves tied up, and trees should be staked and protected from the wind. Asparagus beds should be made at the end of the month. Plant Potatoes, sow Peas ; Asparagus also may be sown in pots.

The temperature of glasshouses should be looked to by closing them about 2 p.m. in the afternoon, and no moisture should be allowed to remain on the foliage.

INDEX.

A.

Acalypha ...	...	...
hispida...	...	... 40
marginata	...	... 40
Wilkesiana	...	... 40
Acacia nilotica	...	... 16
Adansonia ...	...	...
digitata	...	... 17
Adiantum ...	...	... 69
culture of	...	... 70
Agaricum ...	...	...
campestris	...	... 101
Agaves ...	...	... 74
Americana	...	... 74
glauca ...	...	... 74
Albizzia ...	...	...
Lebbek ...	...	... 17
procera	...	... 17
Aliantus glandulosa		... 17
Aloe...	...	... 74
frutescens	...	... 74
vera	...	... 74
Alpina	...	...
nutans...	...	... 82
American Beans	...	... 103
Annuals	...	... 85
Anona	...	...
squamosa	...	... 90
Anthurium ...	...	... 86
Antigonon ...	...	...
insigne	...	... 48
leptopus	...	... 47
Apricot	...	... 97
Aquatics	...	... 79
Araucaria ...	...	...
Bidwillii	...	... 22
Cookii ...	...	... 22
Cunninghamii		... 22
excelsa	...	... 21
Aralia	...	...
digitata	...	... 39
filicifolia	...	... 40
papyrifera	...	... 40
pentaphylla	...	... 40
Areca	...	...
Baueri...	...	... 32
Areca lutescens	...	... 32
Aristolochia	...	...
Brasiliensis	...	... 48
elegans	...	... 48
gigas ...	...	... 48
Aroids	...	... 68
Artichoke ...	...	... 107
Asparagus ...	...	... 102
officinalis	...	... 102
plumosus	...	... 71
plumosus nanus		... 71
Astrocaryum	...	...
mexicana	...	... 32
Azalia	...	... 45

B.

Balanites...	...	... 18
egyptica	...	...
Balcony Gardening		... 86
Bambusa ...	...	...
arundinacea	...	... 40
japonica	...	... 41
nana ...	...	... 41
nigra ...	...	... 41
Banana	...	... 95
Banksia	...	... 47
Bauhinia	...	... 18
aculeata	...	...
purpurea	...	... 18
reticulata	...	... 18
tomentosa	...	... 18
Beaumontia	...	...
grandiflora	...	... 48
Biota	...	...
orientalis	...	... 41
Bougainvillea	...	... 49
glabra ...	...	... 49
spectabilis	...	... 49
Boussingaultia	...	...
baselloides	...	... 54
Broussonetia	...	...
papyrifera	...	... 18
Buddleia ...	...	...
madagascariensis		... 48
Bulbs	...	... 77
Butter Beans	...	... 103

C.

Cæsalpinia	
Bonducilla	19
pulcherrima	19
Calla	81
Callitris	
quadrivalvis	20
Calotropis	
procera	45
Camellia	45
Canna	83
Cape Gooseberry	96
Carica Papaya	90
Caryota	32
Cassia	20
bicapsularis	20
festula	19
Ceratonia	
Siliqua	19
Cercis	
canadensis	19
Siliquastrum ...	18
Cercus	
trigonis	75
Chamædorea, sp	33
Chamærops	
Fortunei	33
Chrysanthemum	84
culture of	83
Cineraria maritima ...	88
Citharexylum	
cimereum	19
Citrus	
Aurantium	91
Limonum	91
medica	92
nobilis	92
Clematis	
Jackmanii	50
Climbing plants	46
culture of	47
Clitoria	
ternatea	49
Coniferæ	20
Conservatories	67
management of ...	66
Cocos	
flexuosa	32
nucifera	33
plumosa	33
Weddeliana	33
Cordia	
Myxa	18
Croton	41
aucubæfolius superbus	41
aucubæfolius giganteus	42
Cryptomeria	
japonica	20
Cryptostegia	
grandiflora	49
Cupressus	
fastigiata	20
horizontalis	20
sempervirens	20
Cycas	
circinalis	37
revoluta	37
Cyperus	
rotundus	11
Cynodon	
dactylon	11

D.

Dahlia	78
Datura	
alba	43
fastuosa	43
rosea	43
suaveolens	43
Dolichos	
sinensis	54
Dracæna	
ferrea	43
culture of	43
Duranta	
Plumieri	14

E.

Egg Plant	104
Epiphyllum	75
Eriobotrya	
japonica	93
Erythrina	
crista-galli	23
Corallodendron ...	24
indica	24
Eucharis	
amazonica	79
Eucalyptus	

Index.

Eucalyptus amygdalina ... 24	Hoya carnosa 50
globulus 24	Hyacinths 78
resinifolia 24	Hyophorbe ... ...
Euphorbiaceæ 75	amaricaulis 34
Euphorbia	Hyphæne ... ...
jacquiniæflora ... 23	thebaica 34
nerifolia 23	I.
pulcherrima 23	Indian Fig 96
pulcherrima alba ... 23	Ipomæa ... ...
F.	Bona-nox 50
Ficus 24	cairica 51
bengalensis 25	Iris ... ...
Carica 94	germanica 81
eriobotryoides 25	J.
infectoria 26	Jacaranda ... ...
japonica 26	mimosifolia 26
macrophylla 25	Jasminum ... ...
nymphæifolia 25	Double Tuscan ... 51
religiosa 26	fruticans 51
retusa var nitida ... 26	grandiflora 51
sycamorus 26	humile 51
Fig 94	officinale 51
Flower Beds 12	revolutum 51
Fragaria vesca 93	Sambac 51
French Beans 103	K.
G.	Kentia ... ...
Garden	Australis 34
management of ... 5	Belmoreana 34
Gardenias 45	Canterburyana ... 34
Gazania	Fosteriana 34
rigens 13	Kigelia ... ...
Gladioli 79	pinnata 27
Glasshouses 65	Kleinia ... ...
Gloriosa	repens 75
superba 78	L.
Grapes 99	Laginaria ... ...
H.	vulgaris 52
Hakea	Lagunaria ... ...
Hedera	Pattersonii 27
Helix 50	Lantana ... ...
Hedges 13	nivea 54
Hibiscus 44	Latania ... ...
mutabilis 44	aurea 34
rosa-sinensis 14	borbonica 34
schizopetala 44	rubra 34
syriacus 44	Lawns 9
Hoya 50	Leaf soil 70

Lemon 91	**P.**
Lilium auratum 77	Palms 29
longifolium 77	*Pandanus* ... ...
Lippia nodiflora 10	*odoratissimus*... ... 36
Lonicera chinensis ... 51	*Veitchii* 37
sempervirens, var. minor 51	*Papyrus* ... ...
M.	*antiquorum* 80
	Parkinsonia ... ...
Mangifera	*aculeata* 28
indica 95	*Passiflora* ... ...
Meila	*cœrulea* 54
Azedarach 27	Constance Elliot ... 54
Melaleuca 47	*edule* 54
Mesembryanthemum ... 76	*quadrangularis* ... 54
cordifolium 76	Paths 7
crystallinum 76	Peas... 103
edule 76	Peach 97
Metrosideros 47	*Periploca* ... ...
Metroxylon...	*lævigata* 52
sagus 35	*Phaseolus* ... ...
Momordica...	*americana* 103
balsamina 51	*Caracalla* 54
Monsteria	*vulgaris* 103
deliciosa 82	*Phœnix* ... ...
Musa	*canariensis* 35
enseti 96	*dactylifera* 30
rosocea... 96	*leonensis* 35
sapientum 95	*rupicola* 35
Mushrooms... 101	*Physalis Peruviana* ... 96
Mysotis	*Pinus* ... ...
palustris 81	*elata* 23
Myrtus communis... ... 14	*halepensis* 22
	longifolia 23
N.	*Pinaster* 23
Narcissus 77	*Pinea* 23
Nelumbium	*Pisum* ... ...
speciosum 80	*sativum* 103
New Holland Plants ... 47	*Pittosporum* ... ...
Nymphœa	*undulatum* 44
cærula... 81	*Plumeria* 28
Lotos 81	*Podocarpus*... ... ...
	Totara... 21
O.	*Poinciana* ... ...
Orange 91	*regia* 27
Orchids 71	Pomegranate 98
Terrestrial 69	Potato 104
Celestial 69	Potting 69
Opuntia	Potting Soil 70
ficus-indica 96	*Prunus* 97

Index.

Prunus armeniaca...	...	97
domestica	... ...	98
Persica...	... ...	97
Punica	... ...	
Granatum	... ...	98
Pyrus	... ...	
communis	... ...	98
Malus...	... ...	98

Q.

Quisqualis...	... ...	
indica	... ...	52

R.

Radish	... ...	
Ravenala	... ...	
madagascariensis	...	83
Rhapis	... ...	
flabelliformis...	...	35
Rhynchospermum...	...	
jasminoides	... ...	52
Richardia (or *Calla*)	..	
æthiopica	... ...	81
maculata	... ...	81
Roses	... ...	56
cultivation	... ...	57
descriptive list	...	58

S.

Sabal	... ...	
Blackburniana	...	35
princeps	... ...	36
Salisburia ...	... ...	
adiantifolia	... ...	21
Salix	... ...	
ægyptica	... ...	28
babylonica	... ...	28
Schinus	... ...	
Molle ...	... ...	28
terebinthifolia	...	14
Seaforthia ...	... ...	
elegans...	... ...	36
Semi-aquatics	... ...	80
Solandra ...	... ...	
grandiflora	... ...	53
Solanum	... ...	
Melongena	... ...	104
Seaforthiana ...	...	53
tuberosum	... ...	104
Wendlandii	... ...	53

Sphagnum Moss	... ...	68
Stephanotis...	... ...	53
floribunda	... ...	52
Strawberry...	... ...	93
Strelitzia ...	... ...	
reginæ ...	... ...	82

T.

Taxodium ...	... ...	
distichum	... ...	20
Taxus	... ...	
baccata...	... ...	21
fastigiata	... ...	21
Tecoma	... ...	
capensis	... ...	53
Cavendishii	... ...	28
jasminioides	... ...	53
radicans	... ...	53
stans	... ...	28
Thalia dealbata	... ...	82
Thrinax ...	... ...	
elegans...	... ...	36
parviflora	... ...	36
Tradescantia	... ...	
tricolor	... ...	86
Tuberose ...	... ...	79
Turnip	... ...	103

V.

Vegetables ...	... ...	101
Victoria	... ...	
Regia ...	... ...	82
Vitis vinifera	... ...	99

W.

Washingtonia	... ...	
filifera ...	... ...	36
Wistaria ...	... ...	
polystachya	... ...	54

X.

Xanthosoma	... ...	
robusta	... ...	83

Y.

Yucca	... ...	76
gloriosa	... ...	76
aloifolia	... ...	76

Z.

Zamia	... ...	37
integrifolia	... ...	37
Lindeni	... ...	37

THE LARGEST AND FINEST STOCK
OF
ORCHIDS
IN THE WORLD.

Establishments: ST. ANDRÉ, BRUGES, and ST. ALBANS, HERTS.

Inspection Invited.

SANDER,
ST. ALBANS.

THE Great Northern Railway Company have opened a station in our establishment (St. Albans), and visitors by giving notice (at Hatfield station) to the Guard of the train they travel by will alight in our glass houses. The establishment is reached in half-an-hour by Midland Railway from St. Pancras, and forty-five minutes by London and North-Western Railway from Euston. The Nurseries are entirely devoted to the cultivation of Orchids and new plants, and contain nearly four acres of glass.

Quotations, Priced Lists for all Climates, and all particulars on application.

PLANTS, SEEDS, & BULBS.

M̲R. WILLIAM BULL'S Illustrated Catalogue of **NEW AND RARE PLANTS**, price 1s., contains names and prices of:

Choice New and Rare Plants	Choice Stove Plants
Choice Greenhouse Plants	Choice Hardy Plants
Choice Ferns	Choice Orchids
Choice Crotons	Choice Liliums
Choice Pelargoniums	Choice Palms
Choice Caladiums	Choice Chrysanthemums

&c., &c.

The CATALOGUE of SEEDS

Contains many novelties of sterling merit both in

VEGETABLE AND FLOWER SEEDS.

THE BULB CATALOGUE

Contains prices of all the best kinds in

HYACINTHS, TULIPS, CROCUS, DAFFODILS and NARCISSUS, GLADIOLUS, IRIS, &c., &c.

APPLY TO

WILLIAM BULL, F.L.S.,

Establishment for New and Rare Plants and Seeds,

536, KING'S ROAD, CHELSEA, LONDON, S.W.

ESTABLISHED 1854.   BY ROYAL WARRANT.   ESTABLISHED 1854.

B. S. WILLIAMS & SON,

Nurserymen, Seedsmen, and Florists to the Queen,

Victoria and Paradise Nurseries,

UPPER HOLLOWAY, LONDON, N.,
AND AT 169, PICCADILLY, LONDON, W.

NURSERY DEPARTMENT.

These Nurseries contain a large show Conservatory, Orchid Houses, new Plant Houses, Fern Houses, Palm Stoves, Greenhouses, &c., filled with handsome specimens. It is at all times well worthy of a visit, and is easy of access.

NEW AND GENERAL PLANT CATALOGUE. (*Published in Spring, post free.*) Containing select Lists of Choice Exotic Orchids, Ferns, Stove, and Greenhouse Plants, Indian Azaleas, Camellias, Variegated and Ornamental Foliaged Plants, Fuchsias, Pelargoniums, Rhododendrons, Verbenas, Petunias, Phloxes, Cinerarias, Chrysanthemums, Hardy Variegated Plants, Roses, &c. A large collection of all the leading as well as all the new varieties of CHRYSANTHEMUMS is always on hand.

SEED AND BULB DEPARTMENT.

Particular attention is given to this branch of the business. B. S. W. and Son's worldrenowned Flower and Vegetable Seeds are selected with great care, warranted true to name, and are unrivalled for excellence of quality.

Potatoes, Agricultural Seeds, Horticultural Implements, Sundries, and all Garden Requisites kept in stock.

Home and Export Orders promptly executed and forwarded by the most expeditious route.

DESCRIPTIVE AND PRICED SEED CATALOGUE. (*Published in January post free.*) All orders for Seeds amounting to £1 and upwards will be delivered Free of Carriage. All Packets of Flower Seeds are forwarded Free by Post.

BULB CATALOGUE. (*Published in Autumn, post free.*) Includes Selected Lists of B. S. W. and Son's Gold Medal Exhibition and other choice Hyacinths, Tulips, Narcissus, Crocus, Ranunculus, Amaryllis, Ixias, Gladioli, Liliums, Begonias, and all other choice and rare kinds of Bulbs of the finest quality, with useful suggestions as to culture, &c.; also

**FINE STOCKS OF VINES, FRUIT TREES, SHRUBS
and TREES of Every Description.**

FURNISHING CONSERVATORIES, &c.

Estimates are also given for keeping Conservatories, Entrance Halls, Jardinieres, Fern Cases, Window Boxes, &c., regularly filled with Plants; also for supplying Plants for Table and Drawing-room Decoration, changing them as often as may be necessary.

Our Bouquets for Weddings and Balls, our Wreaths, Crosses, and other Designs for Funerals, also Loose Cut Flowers, Ferns, &c., for Table Decoration, executed by our Court Florist, are unsurpassed by any House in Town.

Cut Flowers, securely packed, can be sent to any part of the United Kingdom by Parcel Post upon the shortest notice, on receipt of either letter or telegram.

Floral Decorations for Balls, Fêtes. and Public Ceremonies carried out to any extent. A large Stock of Plants and Cut Flowers is kept at the West End Branch, where Orders may be received for any article we supply.

Landscape Gardening and Horticultural Building.

Estimates on the lowest terms are given for laying out Parks, Pleasure Grounds, Gardens, &c., also for erecting Greenhouses and Conservatories, and planting of Ferneries, &c., Heating and Ventilating on the most improved principles. Ladies or Gentlemen who contemplate constructing Conservatories, Stoves, &c., should pay these Nurseries a visit to inspect the Plant Houses, which have been erected by our own Mechanics, when any advice or other information will be gladly given.

THE GARDEN.

Messrs. H. Cannell & Sons solicit orders for whatever is wanted from England. We have the largest and finest Stock, and are the most successful exporters of Plants, Seeds, and Bulbs in the World.

We ask all lovers of "Their Garden" to send for our Catalogues, and also to "Come and See."

E. H. DAY, Esq., Assiout, Upper Egypt, says:

"Dear Sirs,—I am glad that I can get plants and seeds from you in good condition, and although the last lot was received in the hot season, I only lost two. I am coming to England shortly, and will call and order some more to be sent here."

[This gentleman did so, and will no doubt gladly speak of us if applied to.—H. C. & SONS.]

H.I.M. THE SULTAN OF TURKEY.

"Imperial Palace, Bache Naze, Constantinople,
"*March 20th.*

"Have received the plants and seeds according to my order in good condition, and beg to thank you for your kind attention.

"Truly yours,
"His Excellency RAOUF BEY."

And also on May 31st, 1895.

"I herewith send remittance for plants, seeds, &c., and am glad to say they arrived in good condition, and are satisfactory.

"Truly yours,
"His Excellency RAOUF BEY."

H. CANNELL & SONS,
SWANLEY, KENT,
ENGLAND.

SUTTON'S SEEDS
For use in EGYPT.

SUTTON'S
SPECIAL EXPORT COLLECTIONS
OF
VEGETABLE AND FLOWER SEEDS

Supplied at the undermentioned prices:—

VEGETABLE SEEDS.			FLOWER SEEDS.		
£	s.	d.	£	s.	d.
10	10	0	5	5	0
5	5	0	4	4	0
4	4	0	3	3	0
3	3	0	2	2	0
2	2	0	1	11	6
1	1	0	1	1	0
0	10	6	0	10	6
0	7	6	0	7	6
0	5	0	0	5	0

Sutton's Composite Collection of
VEGETABLE AND FLOWER SEEDS, 10s. 6d.

All the prices include an air-tight case, which will be found useful for many purposes after the Seeds have been removed.

PRICED LISTS OF VEGETABLE AND FLOWER SEEDS
POST FREE ON APPLICATION.

SUTTON & SONS, READING, ENGLAND.

CARTERS'
Post Free Collections of
"TESTED" ENGLISH SEEDS.
Carefully Arranged & Suitably Packed for all Parts of the World.

VEGETABLE SEEDS.

PEAS, best sorts.
CAULIFLOWER, best sorts.
KOHL RABI.
ONIONS, best sorts.
COUVE TRONCHUDA.
SPINACH.
BEANS, Dwarf and Runner.
BEET.
CELERY, White and Red.
LEEK.

POTATO SEED.
TURNIP, White and Yellow.
KALE, Thousand-headed.
CABBAGE, best sorts.
LETTUCE, Cos and Cabbage.
PARSLEY.
VEGETABLE MARROW.
BRUSSELS SPROUTS.

SAVOY.
CAPSICUM.
CUCUMBER.
TOMATO.
MELON.
RADISH, best sorts.
CARROT, best sorts.
HERBS.
EGG PLANT.
MUSTARD AND CRESS.

Carters' Box, containing useful quantities of the above varieties, sent packing and postage free to all parts within the Parcel Post Union, on receipt of remittance or an order for payment in England, price 16s.

Boxes containing larger quantities, price 23s., 43s., 56s., 70s., 90s., and 120s.

FLOWER SEEDS.

ASTER.
AMARANTHUS.
ANTIRRHINUM.
BALSAM.
BEGONIA.
CALLIOPSIS.
CANARY CREEPER.
CANDYTUFT.
CELOSIA.
CHRYSANTHEMUM.
COBÆA.
CONVOLVULUS.
DAHLIA.
DELPHINIUM.
DIANTHUS.
ESCHSCHOLTZIA.

EVERLASTING FLOWER.
GAILLARDIA.
GLOBE AMARANTHUS.
GODETIA.
GOLDEN FEATHER.
HEARTSEASE.
HELIOTROPE.
IPOMÆA.
LARKSPUR.
LINUM.
LOBELIA.
LUPINUS.
MARIGOLD.
MARVEL OF PERU.
MAURANDYA.
MESEMBRYANTHEMUM.

MIGNONETTE.
NASTURTIUM.
NEMOPHILA.
PERILLA.
PETUNIA.
PHLOX.
POLYANTHUS.
PORTULACA.
SALPIGLOSSIS.
STOCK.
SWEET PEA.
SWEET WILLIAM.
VERBENA.
WALLFLOWER.
ZEA.
ZINNIA.

Carters' Box, containing large packets of the above, and giving an infinite variety of colour, sent packing and postage free to all parts within the Parcel Post Union, on receipt of remittance or an order for payment in England. price 22s.

Smaller Boxes, containing selections from the above list, price 7s. 6d., 11s., and 17s.

All Orders must be accompanied by a Remittance to Cover Value, or the Goods must be paid for in England.

Postal Orders to be made payable to JAMES CARTER & CO., at the General Post Office, London.

WE SHALL BE PLEASED TO SEND COPIES OF OUR
Illustrated Catalogues of Agricultural or Garden Seeds
Post Free to any part of the world on application.

Seedsmen by Royal Warrants to H.M. the Queen, and H.R.H. the Prince of Wales, to the Home, Colonial, and many Foreign Governments.

237, 238, & 97, HIGH HOLBORN, LONDON.

Ransomes'

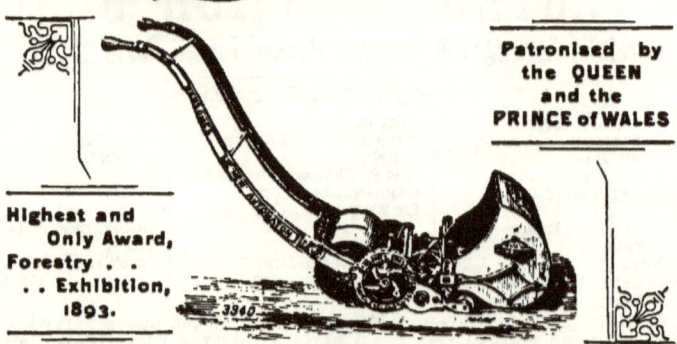

Patronised by the QUEEN and the PRINCE of WALES

Highest and Only Award, Forestry .. .. Exhibition, 1893.

Lawn Mowers.

RANSOMES' "NEW AUTOMATON,"
WITH CHAIN OR WHEEL GEARING.

Confidently recommended as **THE BEST LAWN MOWER IN THE WORLD**. Graceful in design, easy to work, quiet in action, cuts without ribbing, producing a fine, velvety surface. Rolls the whole of the Lawn; no further rolling required. Has a large, open Cylinder, dealing with long or short grass with equal facility; this Cylinder is adjusted by a Patent Single Screw Adjustment which cannot get out of order. The Gearing is simple, quiet and certain; is completely covered, and allows free motion to the knives. Has the best materials and workmanship, and highest finish. Will give satisfaction for years, and can be easily repaired.

	8in.	10in.	12in.	14in.	16in.	18in.	20in.	22in.	24in.
PRICES:—	55/-	70/-	90/-	110/-	130/-	150/-	170/-	190/-	210/-
Packing for Export extra.	6/-	6/-	7/-	7/-	8/-	8/-	8/-	10/-	10/-

The prices given above include delivery to the English Shipping Port. Packing charges 1/- less if without Grass Box.

RANSOMES' "ANGLO-PARIS" MOWER.
The best Light Machine.

Suitable for Small Gardens and for Ladies' and Amateurs' use. Has perfect adjustment, and is unequalled by any similar machine in the market.

Complete Illustrated Catalogues Free
ON APPLICATION TO—

RANSOMES, SIMS & JEFFERIES, LIMITED,
IPSWICH, ENGLAND.

W. RICHARDSON & CO.,

MANUFACTURERS OF

Horticultural Buildings,

Conservatories,

Greenhouses,

Vineries,

PEACH HOUSES, STOVE AND ORCHID HOUSES, &c.,

Fitted with the best up-to-date improvements and in keeping with adjacent buildings.

Awarded the only **GOLD MEDAL** for Horticultural Buildings
At the International Horticultural Exhibition, London.
AND MANY OTHER PRIZE MEDALS.

PARISIAN GREENHOUSE LATH BLINDS,
AWARDED OVER 50 MEDALS.

Any house made portable for exportation. All the parts numbered, so that unskilled labourer can fix. Glass cut to exact sizes, and everything supplied to make the house complete. Carriage paid to any British Port.

IMPROVED SPAN GARDEN FRAMES
(with Adjustable Ridge Ventilation).

Contractors for Steam and Hot Water Heating Apparatus, &c.
Efficiency Guaranteed, at the lowest possible Prices for good work.

New Catalogue—The most Artistic & Complete Issued, Free.

W. RICHARDSON & CO., DARLINGTON.

SEEDS AND BULBS FOR ALL CLIMATES AND SEASONS.

BARR'S SUPERIOR SEEDS
FOR FLOWER & KITCHEN GARDEN

BARR'S   DESCRIPTIVE CATALOGUE OF
(Ready 1st January.)   **BEST VEGETABLE SEEDS**
For Sowing the Year Round.

BARR'S COLLECTIONS OF BEST VEGETABLE SEEDS,
12 6, 21 -, 42 -, 63 -, and upwards.

BARR'S   DESCRIPTIVE CATALOGUE OF
(Ready 1st January.)   **BEST FLOWER SEEDS**
For Out of Doors and Under Glass.

BARR'S COLLECTIONS OF FLOWER SEEDS FOR ABROAD, 5 6, 7 6, 10 6, 21 -, 42 -, and 63 -.

BARR'S   DESCRIPTIVE CATALOGUE OF
(Ready 1st September.)   **BEST FLOWERING BULBS**
For Flower Garden and Early Forcing.

BARR'S COLLECTIONS of BULBS, 12 6, 21 -, 42 -, & 63 -.

BARR'S   DESCRIPTIVE CATALOGUE OF
(Ready in August.)   **NEW ENGLISH DAFFODILS**
For Flower Garden, Indoors, and Naturalization.

BARR'S COLLECTIONS OF ENGLISH DAFFODILS,
7 6, 10 6, 15 - 21 -, 42 -, 63 -, and 84 -.

☞ ALL CATALOGUES FREE ON APPLICATION.

BARR AND SON,
12 and 13, KING STREET,
COVENT GARDEN, LONDON.

NURSERIES AT LONG DITTON, SURREY.

Cable Address: *"WARE, TOTTENHAM, LONDON."*

CHRYSANTHEMUMS.
All the various sections represented by the most worthy varieties.

For the Open, and for Greenhouse cultivation.

CATALOGUE READY IN FEBRUARY.

HARDY PERENNIALS
In immense quantities.

Easy to grow, and adapted for every conceivable outdoor position.

CATALOGUE ISSUED IN FEBRUARY.

TUBEROUS BEGONIAS.
Undoubtedly the very finest strain procurable.

THE HIGHEST AWARDS gained at all the principal Exhibitions, including Three Gold Medals, and numerous Silver Medals and First Prizes.

CATALOGUE READY IN JAN.

WARE'S

FLORISTS' FLOWERS.
Comprising CARNATIONS and PICOTEES by Hundreds of Thousands, in the best sorts only. Also Collections of PHLOXES, PYRETHRUMS, POTENTILLAS, DELPHINIUMS, PENTSTEMONS, PANSIES, VIOLETS, &c.

CATALOGUE IN FEBRUARY.

DAHLIAS.
CACTUS DAHLIAS of the finest kinds ever raised. Single, Pompone, and Show varieties in all the best sorts.

CATALOGUE devoted to these and other plants for summer Bedding, viz., Marguerites, Salvias, &c., SENT OUT IN APRIL.

VEGETABLE AND FLOWER SEEDS.
Strains of reliable and proved worth, tested and true. Complete Collections of SELECT VEGETABLE SEEDS and CHOICE FLOWER SEEDS for Small, Medium, and Large Gardens.

CATALOGUE PUBLISHED ANNUALLY IN JANUARY.

SPECIALITIES.

NARCISSUS AND DAFFODILS.
Healthy, sound Bulbs of these charming Spring flowers, in prodigious numbers & bewildering variety.

LILIES.
The finest and most complete Collection in the Country.

IRISES.
A most unique Collection, probably the largest in existence; every known species and variety worth cultivation being grown.

HYACINTHS, TULIPS, CROCUSES, OFFERED IN AUTUMN CATALOGUE OF BULBS.

GLADIOLUS; PÆONIES (Tree and Herbaceous); HARDY FERNS; PRIMULAS; ROSES; CLIMBERS and CREEPING PLANTS; AQUATIC and BOG PLANTS; ROCKERY and BORDER PLANTS. Decorative Shrubs, &c.

Selections of Plants recommended for any kind of Soil or Situation upon application.

Copies of any and all of above Catalogues may be had Free on application.

(KINDLY MENTION THIS BOOK WHEN APPLYING FOR CATALOGUES.)

THOMAS S. WARE,
HALE FARM NURSERIES
TOTTENHAM, LONDON.

CLAY'S LONDON FERTILIZER

TRADE MARK.

UNSURPASSED

FOR

CHRYSANTHEMUMS, VINES, ROSES,

AND

ALL HORTICULTURAL PURPOSES.

They are used by the leading Growers, R.B.S., R.H.S., Royal Parks, Kew Gardens, London County Council, throughout the United Kingdom, and in every quarter of the Globe.

For all pot Plants Clay's "Fertilizer" should be used, as the Plants require manure to bring them to perfection in the Egyptian climate.

"I know of no chemical manure to give better results.
   "WALTER DRAPER,
    "F.R.H.S."

Crushed Bones, Peruvian Guano, Sulphate of Ammonia, Nitrate of Soda, and other Manures, Tobacco Cloth and Paper.

Best qualities only.

Prices on application.

Clay's Manures

PRODUCTIVE. SAFE. LASTING. ECONOMICAL.

The respective Trade Mark is printed on every Packet and Bag, and impressed on the Lead Seal attached to the mouth of each Bag, which is THE ONLY GUARANTEE OF GENUINENESS.

Sold by Seedsmen, Florists, and Nurserymen in 6d. and 1s. Packets, and SEALED BAGS: 7lbs. 2s. 6d.; 14lbs. 4s. 6d.; 28lbs. 7s. 6d.; 56lbs. 12s. 6d.; 112lbs. 20s. Or direct from the Works, Carriage paid in the United Kingdom for Cash with Order (except 6d. Packets).

"There is nothing grown on the land, whether of Fruit, Flowers, or Vegetables, that will not benefit by its judicious use.
   "SHIRLEY HIBBERD.
    "F.R.H.S."

"I am really surprised at its effects, which are such as I have never seen follow the use of anything else in the shape of manure.
   "THOMAS BAINES,
    "F.R.H.S."

For Palms, Crotons. Aroids, Bulbs, Ferns, and all choice Plants, a small teaspoonful sprinkled over the soil and watered in will be found very beneficial.

CLAY & SON,

Manure Manufacturers, . . .
. . Bone Crushers, &c.,

TEMPLE MILL LANE, STRATFORD, LONDON, E.

TRADE MARK.

Catalogue of Practical Handbooks Published by L. Upcott Gill, 170, Strand, London, W.C.

ANGLER, BOOK OF THE ALL-ROUND. A Comprehensive Treatise on Angling in both Fresh and Salt Water. In Four Divisions, as named below. By JOHN BICKERDYKE. With over 220 Engravings *In cloth, price 5s. 6d., by post 6s.* (A few copies of a LARGE PAPER EDITION, *bound in Roxburghe, price 25s.*)

    Angling for Coarse Fish. Bottom Fishing, according to the Methods in use on the Thames, Trent, Norfolk Broads, and elsewhere. Illustrated. *In paper, price 1s., by post 1s. 2d.*

    Angling for Pike. The most Approved Methods of Fishing for Pike or Jack. Profusely Illustrated. *In paper, price 1s., by post 1s. 2d.; cloth, 2s. (uncut), by post 2s. 3d.*

    Angling for Game Fish. The Various Methods of Fishing for Salmon; Moorland, Chalk-stream, and Thames Trout; Grayling and Char. Well Illustrated. *In paper, price 1s. 6d., by post 1s. 9d.*

    Angling in Salt Water. Sea Fishing with Rod and Line, from the Shore, Piers, Jetties, Rocks, and from Boats; together with Some Account of Hand-Lining. Over 50 Engravings. *In paper, price 1s., by post 1s. 2d.; cloth, 2s. (uncut), by post 2s. 3d.*

AQUARIA, BOOK OF. A Practical Guide to the Construction, Arrangement, and Management of Fresh-water and Marine Aquaria; containing Full Information as to the Plants, Weeds, Fish, Molluscs, Insects, &c., How and Where to Obtain Them, and How to Keep Them in Health. Illustrated. By REV. GREGORY C. BATEMAN, A.K.C., and REJINALD A. R. BENNETT, B.A. *In cloth gilt, price 5s. 6d., by post 5s. 10d.*

AQUARIA, FRESHWATER: Their Construction, Arrangement, Stocking, and Management. Fully Illustrated. By REV. G. C. BATEMAN, A.K.C. *In cloth gilt, price 3s. 6d., by post 3s. 10d.*

AQUARIA, MARINE: Their Construction, Arrangement, and Management. Fully Illustrated. By R. A. R. BENNETT, B.A. *In cloth gilt, price 2s. 6d., by post 2s. 9d.*

AUSTRALIA, SHALL I TRY? A Guide to the Australian Colonies for the Emigrant Settler and Business Man. With two Illustrations. By GEORGE LACON JAMES. *In cloth gilt, price 3s. 6d., by post 3s. 10d.*

AUTOGRAPH COLLECTING: A Practical Manual for Amateurs and Historical Students, containing ample information on the Selection and Arrangement of Autographs, the Detection of Forged Specimens, &c., &c., to which are added numerous Facsimiles for Study and Reference, and an extensive Valuation Table of Autographs worth Collecting. By HENRY T. SCOTT, M.D., L.R.C.P., &c., Rector of Swettenham, Cheshire. *In leatherette gilt, price 7s. 6d., by post 7s. 10d.*

BEES AND BEE-KEEPING: Scientific and Practical. By F. R. CHESHIRE, F.L.S., F.R.M.S., Lecturer on Apiculture at South Kensington. *In two vols., cloth gilt, price 16s., by post 16s. 4d.*

Vol. I., Scientific. A complete Treatise on the Anatomy and Physiology of the Hive Bee. *In cloth gilt, price 7s. 6d., by post 7s. 10d.*

Vol. II., Practical Management of Bees. An Exhaustive Treatise on Advanced Bee Culture. *In cloth gilt, price 8s. 6d., by post 8s. 10d.*

BEE-KEEPING, BOOK OF. A very practical and Complete Manual on the Proper Management of Bees, especially written for Beginners and Amateurs who have but a few Hives. Fully Illustrated. By W. B. WEBSTER, First-class Expert, B.B.K.A. *In paper, price 1s., by post 1s. 2d.; cloth, 1s. 6d., by post 1s. 8d.*

BEGONIA CULTURE, for Amateurs and Professionals. Containing Full Directions for the Successful Cultivation of the Begonia, under Glass and in the Open Air. Illustrated. By B. C. RAVENSCROFT. *In paper, price 1s., by post 1s. 2d.*

BENT IRON WORK: A Practical Manual of Instruction for Amateurs in the Art and Craft of Making and Ornamenting Light Articles in imitation of the beautiful Mediæval and Italian Wrought Iron Work. By F. J. ERSKINE. Illustrated. *In paper, price 1s., by post 1s. 2d.*

BOAT BUILDING AND SAILING, PRACTICAL. Containing Full Instructions for Designing and Building Punts, Skiffs, Canoes, Sailing Boats, &c. Particulars of the most suitable Sailing Boats and Yachts for Amateurs, and Instructions for their Proper Handling. Fully Illustrated with Designs and Working Diagrams. By ADRIAN NEISON, C.E., DIXON KEMP, A.I.N.A., and G. CHRISTOPHER DAVIES. *In one vol., cloth gilt, price 7s. 6d., by post 7s. 10d.*

BOAT BUILDING FOR AMATEURS, PRACTICAL. Containing Full Instructions for Designing and Building Punts, Skiffs, Canoes, Sailing Boats, &c. Fully Illustrated with Working Diagrams. By ADRIAN NEISON, C.E. Second Edition, Revised and Enlarged by DIXON KEMP, Author of "Yacht Designing," "A Manual of Yacht and Boat Sailing," &c. *In cloth gilt, price 2s. 6d., by post 2s. 9d.*

BOAT SAILING FOR AMATEURS. Containing Particulars of the most Suitable Sailing Boats and Yachts for Amateurs, and Instructions for their Proper Handling, &c. Illustrated with numerous Diagrams. By G. CHRISTOPHER DAVIES. Second Edition, Revised and Enlarged, and with several New Plans of Yachts. *In cloth gilt, price 5s., by post 5s. 4d.*

BOOKBINDING FOR AMATEURS: Being Descriptions of the various Tools and Appliances Required, and Minute Instructions for their Effective Use. By W. J. E. CRANE. Illustrated with 156 Engravings. *In cloth gilt, price 2s. 6d., by post 2s. 9d.*

BUNKUM ENTERTAINMENTS: A Collection of Original Laughable Skits on Conjuring, Physiognomy, Juggling, Performing Fleas, Waxworks, Panorama, Phrenology, Phonograph, Second Sight, Lightning Calculators, Ventriloquism, Spiritualism, &c., to which are added Humorous Sketches, Whimsical Recitals, and Drawing-room Comedies. *In cloth, price 2s. 6d., by post 2s. 9d.*

BUTTERFLIES, THE BOOK OF BRITISH: A Practical Manual for Collectors and Naturalists. Splendidly Illustrated throughout with very accurate Engravings of the Caterpillars, Chrysalids, and Butterflies, both upper and under sides, from drawings by the Author or direct from Nature. By W. J. LUCAS, B.A. *Price 3s. 6d., by post 3s. 9d.*

BUTTERFLY AND MOTH COLLECTING: Where to Search, and What to Do. By G. E. SIMMS. Illustrated. *In paper, price 1s., by post 1s. 2d.*

CACTUS CULTURE FOR AMATEURS: Being Descriptions of the various Cactuses grown in this country; with Full and Practical Instructions for their Successful Cultivation. By W. WATSON, Assistant Curator of the Royal Botanic Gardens, Kew. Profusely Illustrated. *In cloth gilt, price 5s., by post 5s. 3d.*

CAGE BIRDS, DISEASES OF: Their Causes, Symptoms, and Treatment. A Handbook for everyone who keeps a Bird. By DR. W. T. GREENE, F.Z.S. *In paper, price 1s., by post 1s. 2d.*

CANARY BOOK. The Breeding, Rearing, and Management of all Varieties of Canaries and Canary Mules, and all other matters connected with this Fancy. By ROBERT L. WALLACE. Third Edition. *In cloth gilt, price 5s., by post 5s. 4d.; with COLOURED PLATES, 6s. 6d., by post 6s. 10d.;* and as follows:

General Management of Canaries. Cages and Cage-making Breeding, Managing, Mule Breeding, Diseases and their Treatment, Moulting, Pests, &c. Illustrated. *In cloth, price 2s. 6d., by post 2s. 9d.*

Exhibition Canaries. Full Particulars of all the different Varieties, their Points of Excellence, Preparing Birds for Exhibition, Formation and Management of Canary Societies and Exhibitions. Illustrated. *In cloth, price 2s. 6d., by post 2s. 9d.*

CANOE BUILDING FOR AMATEURS: A Practical Manual, with Plans, Working Diagrams, and full Instructions. By COTTERILL SCHOLEFIELD. *Price 2s. 6d., by post 2s. 9d.* [*In the Press.*

CARD TRICKS, BOOK OF, for Drawing-room and Stage Entertainments by Amateurs; with an exposure of Tricks as practised by Card Sharpers and Swindlers. Numerous Illustrations. By PROF. R. KUNARD. *In illustrated wrapper, price 2s. 6d., by post 2s. 9d.*

CATS, DOMESTIC OR FANCY: A Practical Treatise on their Antiquity, Domestication, Varieties, Breeding, Management, Diseases and Remedies, Exhibition and Judging. By JOHN JENNINGS. Illustrated. *In cloth, price 2s. 6d., by post 2s. 9d.*

CHRYSANTHEMUM CULTURE, for Amateurs and Professionals. Containing Full Directions for the Successful Cultivation of the Chrysanthemum for Exhibition and the Market. Illustrated. By B. C. RAVENSCROFT. *In paper, price 1s., by post 1s. 2d.*

COINS, A GUIDE TO ENGLISH PATTERN, in Gold, Silver, Copper, and Pewter, from Edward I. to Victoria, with their Value. By the REV. G. F. CROWTHER, M.A. Illustrated. *In silver cloth, with gilt facsimiles of Coins, price 5s., by post 5s. 3d.*

COINS OF GREAT BRITAIN AND IRELAND, A GUIDE TO THE, in Gold, Silver and Copper, from the Earliest Period to the Present Time, with their Value. By the late Colonel W. STEWART THORBURN. With 32 Plates in Gold, Silver, Copper, &c. *In cloth gilt, price 7s. 6d., by post 7s. 10d.*

COLLIE, THE. Its History, Points, and Breeding. By HUGH DALZIEL. Illustrated with Coloured Frontispiece and Plates. *In paper, price 1s., by post 1s. 2d.; cloth, 2s., by post 2s. 3d.*

COLLIE STUD BOOK. Edited by HUGH DALZIEL. *Price* 3s. 6d. *each by post* 3s. 9d. *each.*

Vol. I., containing Pedigrees of 1308 of the best-known Dogs, traced to their most remote known ancestors; Show Record to Feb., 1890, &c.

Vol. II. Pedigrees of 795 Dogs, Show Record, &c.

Vol. III. Pedigrees of 786 Dogs, Show Record, &c.

COLUMBARIUM, MOORE'S. Reprinted Verbatim from the original Edition of 1735, with a Brief Notice of the Author. By W. B. TEGETMEIER, F.Z.S., Member of the British Ornithologists' Union. *Price* 1s., *by post* 1s. 2d.

CONJURING, BOOK OF MODERN. A Practical Guide to Drawing-room and Stage Magic for Amateurs. By PROFESSOR R. KUNARD. Illustrated. *In illustrated wrapper, price* 2s. 6d., *by post* 2s. 9d.

COOKERY FOR AMATEURS; or, French Dishes for English Homes of all Classes. Includes Simple Cookery, Middle-class Cookery, Superior Cookery, Cookery for Invalids, and Breakfast and Luncheon Cookery. By MADAME VALÉRIE. Second Edition. *In paper, price* 1s., *by post* 1s. 2d.

CUCUMBER CULTURE FOR AMATEURS. Including also Melons, Vegetable Marrows, and Gourds. Illustrated. By W. J. MAY. *In paper, price* 1s., *by post* 1s. 2d.

CYCLIST'S ROUTE MAP of England and Wales. The Third Edition; thoroughly Revised. Shows clearly all the Main, and most of the Cross, Roads, and the Distances between the Chief Towns, as well as the Mileage from London. In addition to this, Routes of *Thirty of the most Interesting Tours* are printed in red. The map is mounted on linen, and is the fullest, handiest, and best tourist's map in the market. *In cloth, price* 1s., *by post* 1s. 2d.

DOGS, BREAKING AND TRAINING: Being Concise Directions for the proper education of Dogs, both for the Field and for Companions. Second Edition. By "PATHFINDER." With Chapters by HUGH DALZIEL. Illustrated. *In cloth gilt, price* 6s. 6d., *by post* 6s. 10d.

DOGS, BRITISH, ANCIENT AND MODERN: Their Varieties, History, and Characteristics. By HUGH DALZIEL, assisted by Eminent Fanciers. SECOND EDITION, Revised and Enlarged. Illustrated with First-class COLOURED PLATES and full-page Engravings of Dogs of the Day. This is the fullest work on the various breeds of dogs kept in England. In three volumes, *demy* 8vo, *cloth gilt, price* 10s. 6d. *each, by post* 11s. 1d. *each.*

Dogs Used in Field Sports. Containing Particulars of the following among other Breeds: Greyhound, Irish Wolfhound, Bloodhound, Foxhound, Harrier, Basset, Dachshund, Pointer, Setters, Spaniels, and Retrievers. SEVEN COLOURED PLATES and 21 full-page Engravings.

Dogs Useful to Man in other Work than Field Sports; House and Toy Dogs. Containing Particulars of the following, among other Breeds: Collie, Bulldog, Mastiff, St. Bernards, Newfoundland, Great Dane, Fox and all other Terriers, King Charles and Blenheim Spaniels, Pug, Pomeranian, Poodle, Italian Greyhound, Toy Dogs, &c., &c. COLOURED PLATES and full-page Engravings.

Practical Kennel Management: A Complete Treatise on all Matters relating to the Proper Management of Dogs, whether kept for the Show Bench, for the Field, or for Companions. Illustrated with Coloured and numerous other Plates. [*In the Press.*

DOGS, DISEASES OF: Their Causes, Symptoms, and Treatment; Modes of Administering Medicines; Treatment in cases of Poisoning, &c. For the use of Amateurs. By HUGH DALZIEL. Third Edition. *In paper, price 1s., by post 1s. 2d.; in cloth gilt, 2s., by post 2s. 3d.*

ENTERTAINMENTS, AMATEUR, FOR CHARITABLE AND OTHER OBJECTS: How to Organize and Work them with Profit and Success. By ROBERT GANTHONY. *In coloured cover, price 1s., by post 1s. 2d.*

FANCY WORK SERIES, ARTISTIC. A Series of Illustrated Manuals on Artistic and Popular Fancy Work of various kinds. Each number is complete in itself, and issued at the uniform *price of 6d., by post 7d.* Now ready—(1) MACRAMÉ LACE (Second Edition); (2) PATCHWORK; (3) TATTING; (4) CREWEL WORK; (5) APPLIQUÉ; (6) FANCY NETTING.

FERNS, THE BOOK OF CHOICE: for the Garden, Conservatory, and Stove. Describing the best and most striking Ferns and Selaginellas, and giving explicit directions for their Cultivation, the formation of Rockeries, the arrangement of Ferneries, &c. By GEORGE SCHNEIDER. With numerous Coloured Plates and other Illustrations. *In 3 vols., large post 4to. Cloth gilt, price £3 3s., by post £3 6s.*

FERNS, CHOICE BRITISH. Descriptive of the most beautiful Variations from the common forms, and their Culture. By C. T. DRUERY, F.L.S. Very accurate PLATES, and other Illustrations. *In cloth gilt, price 2s. 6d., by post 2s. 9d.*

FERRETS AND FERRETING. Containing Instructions for the Breeding, Management, and Working of Ferrets. Second Edition, Rewritten and greatly Enlarged. Illustrated. *In paper, price 6d., by post 7d.*

FERTILITY OF EGGS CERTIFICATE. These are Forms of Guarantee given by the Sellers to the Buyers of Eggs for Hatching, undertaking to refund value of any unfertile eggs, or to replace them with good ones. Very valuable to sellers of eggs, as they induce purchases. *In books, with counterfoils, price 6d., by post 7d.*

FIREWORK-MAKING FOR AMATEURS. A complete, accurate, and easily-understood work on Making Simple and High-class Fireworks. By Dr. W. H. BROWNE, M.A. *In paper, price 2s. 6d., by post 2s. 9d.*

FOREIGN BIRDS, FAVOURITE, for Cages and Aviaries. How to Keep them in Health. Fully Illustrated. By W. T. GREENE, M.A., M.D., F.Z.S., &c. *In cloth, price 2s. 6d., by post 2s. 9d.*

FOX TERRIER, THE. Its History, Points, Breeding, Rearing, Preparing, for Exhibition, and Coursing. By HUGH DALZIEL. Illustrated with Coloured Frontispiece and Plates. *In paper, price 1s., by post 1s. 2d.; cloth, 2s., by post 2s. 3d.*

FOX TERRIER STUD BOOK. Edited by HUGH DALZIEL. *Price 3s. 6d. each., by post 3s. 9d. each.*

    Vol. I., containing Pedigrees of over 1400 of the best-known Dogs, traced to their most remote known ancestors.

    Vol. II. Pedigrees of 1544 Dogs, Show Record, &c.

    Vol. III. Pedigrees of 1214 Dogs, Show Record, &c.

    Vol. IV. Pedigrees of 1168 Dogs, Show Record, &c.

    Vol. V. Pedigrees of 1662 Dogs, Show Record, &c.

FRETWORK AND MARQUETRY. A Practical Manual of Instructions in the Art of Fret-cutting and Marquetry Work. Profusely Illustrated. By D. DENNING. *In cloth, price 2s. 6d., by post 2s. 10d.*

FRIESLAND MERES, A CRUISE ON THE. By ERNEST R. SUFFLING. Illustrated. *In paper, price 1s., by post 1s. 2d.*

GAME AND GAME SHOOTING, NOTES ON. Grouse, Partridges, Pheasants, Hares, Rabbits, Quails, Woodcocks, Snipe, and Rooks. By J. J. MANLEY. Illustrated. *In cloth gilt, price 7s. 6d., by post 7s. 10d.*

GAME PRESERVING, PRACTICAL. Containing the fullest Directions for Rearing and Preserving both Winged and Ground Game, and Destroying Vermin; with other Information of Value to the Game Preserver. Illustrated. By WILLIAM CARNEGIE. *In cloth gilt, demy 8vo, price 21s., by post 21s. 9d.*

GARDENING, DICTIONARY OF. A Practical Encyclopædia of Horticulture, for Amateurs and Professionals. Illustrated with 2440 Engravings. Edited by G. NICHOLSON, Curator of the Royal Botanic Gardens, Kew; assisted by Prof. Trail, M.D., Rev. P. W. Myles, B.A., F.L.S., W. Watson, J. Garrett, and other Specialists. *In 4 vols., large post 4to. In cloth gilt, price £3, by post £3 3s.*

GOAT, BOOK OF THE. Containing Full Particulars of the various Breeds of Goats, and their Profitable Management. With many Plates. By H. STEPHEN HOLMES PEGLER. Third Edition, with Engravings and Coloured Frontispiece. *In cloth gilt, price 4s. 6d., by post 4s. 10d.*

GOAT-KEEPING FOR AMATEURS: Being the Practical Management of Goats for Milking Purposes. Abridged from "The Book of the Goat." Illustrated. *In paper, price 1s., by post 1s. 2d.*

GRAPE GROWING FOR AMATEURS. A Thoroughly Practical Book on Successful Vine Culture. By E. MOLYNEUX. Illustrated. *In paper, price 1s., by post 1s. 2d.*

GREENHOUSE MANAGEMENT FOR AMATEURS. The Best Greenhouses and Frames, and How to Build and Heat them, Illustrated Descriptions of the most suitable Plants, with general and Special Cultural Directions, and all necessary information for the Guidance of the Amateur. Second Edition, Revised and Enlarged. Magnificently Illustrated. By W. J. MAY. *In cloth gilt, price 5s., by post 5s. 4d.*

GREYHOUND, THE: Its History, Points, Breeding, Rearing, Training, and Running. By HUGH DALZIEL. With Coloured Frontispiece. *In cloth gilt, demy 8vo, price 2s. 6d., by post 2s. 9d.*

GUINEA PIG, THE, for Food, Fur, and Fancy. Illustrated with Coloured Frontispiece and Engravings. An exhaustive book on the Varieties of the Guinea Pig, and its Management. By C. CUMBERLAND, F.Z.S. *In cloth gilt, price 2s. 6d., by post 2s. 9d.*

HAND CAMERA MANUAL, THE. A Practical Handbook on all Matters connected with the Use of the Hand Camera in Photography. Illustrated. By W. D. WELFORD. Second Edition. *Price 1s., by post 1s. 2d.*

HANDWRITING, CHARACTER INDICATED BY. With Illustrations in Support of the Theories advanced taken from Autograph Letters of Statesmen, Lawyers, Soldiers, Ecclesiastics, Authors, Poets, Musicians, Actors, and other persons. Second Edition. By R. BAUGHAN. *In cloth gilt, price 2s. 6d., by post 2s. 9d.*

170, *Strand, London, W.C.* 7

HARDY PERENNIALS and Old-fashioned Garden Flowers. Descriptions, alphabetically arranged, of the most desirable Plants for Borders, Rockeries, and Shrubberies, including Foliage as well as Flowering Plants. Profusely Illustrated. By J. WOOD. *In cloth, price 5s., by post 5s. 4d.*

HAWK MOTHS, BOOK OF BRITISH. A Popular and Practical Manual for all Lepidopterists. Copiously illustrated in both colours and black and white from drawings from Nature by the Author. By W. J. LUCAS, B.A. *[In the Press.*

HOME MEDICINE AND SURGERY: A Dictionary of Diseases and Accidents, and their proper Home Treatment. For Family Use. By W. J. MACKENZIE, M.D. Illustrated. *In cloth, price 2s. 6d., by post 2s. 9d.*

HORSE-KEEPER, THE PRACTICAL. By GEORGE FLEMING, C.B., LL.D., F.R.C.V.S., late Principal Veterinary Surgeon to the British Army, and Ex-President of the Royal College of Veterinary Surgeons. *In cloth, price 3s. 6d., by post 3s. 10d.*

HORSE-KEEPING FOR AMATEURS. A Practical Manual on the Management of Horses, for the guidance of those who keep one or two for their personal use. By FOX RUSSELL. *In paper, price 1s., by post 1s. 2d.; cloth, 2s., by post 2s. 3d.*

HORSES, DISEASES OF: Their Causes, Symptoms, and Treatment. For the use of Amateurs. By HUGH DALZIEL. *In paper, price 1s., by post 1s. 2d.; cloth 2s., by post 2s. 3d.*

INLAND WATERING PLACES. A Description of the Spas of Great Britain and Ireland, their Mineral Waters, and their Medicinal Value, and the attractions which they offer to Invalids and other Visitors. Profusely illustrated. A Companion Volume to "Seaside Watering Places." *In cloth, price 2s. 6d., by post 2s. 10d.*

JOURNALISM, PRACTICAL: How to Enter Thereon and Succeed. A book for all who think of "writing for the Press." By JOHN DAWSON. *In cloth gilt, price 2s. 6d., by post 2s. 9d.*

LAYING HENS, HOW TO KEEP and to Rear Chickens in Large or Small Numbers, in Absolute Confinement, with Perfect Success. By MAJOR G. F. MORANT. *In paper, price 6d., by post 7d.*

LIBRARY MANUAL, THE. A Guide to the Formation of a Library, and the Values of Rare and Standard Books. By J. H. SLATER, Barrister-at-Law. Third Edition. Revised and Greatly Enlarged. *In cloth gilt, price 7s. 6d., by post 7s. 10d.*

MICE, FANCY: Their Varieties, Management, and Breeding. Re-issue, with Criticisms and Notes by DR. CARTER BLAKE. Illustrated. *In paper, price 6d., by post 7d.*

MILLINERY, HANDBOOK OF. A Practical Manual of Instruction for Ladies. Illustrated. By MME. ROSÉE, Court Milliner, Principal of the School of Millinery. *In paper, price 1s., by post 1s. 2d.*

MODEL YACHTS AND BOATS: Their Designing, Making, and Sailing. Illustrated with 118 Designs and Working Diagrams. A splendid book for boys and others interested in making and rigging toy boats for sailing. It is the best book on the subject now published. By J. DU V. GROSVENOR. *In leatherette, price 5s., by post 5s. 3d.*

MONKEYS, PET, and How to Manage Them. Illustrated. By ARTHUR PATTERSON. *In cloth gilt, price 2s. 6d., by post 2s. 9d.*

MUSHROOM CULTURE FOR AMATEURS. With Full Directions for Successful Growth in Houses, Sheds, Cellars, and Pots, on Shelves, and Out of Doors. Illustrated. By W. J. MAY. *In paper, price 1s, by post 1s. 2d.*

NATURAL HISTORY SKETCHES among the Carnivora—Wild and Domesticated; with Observations on their Habits and Mental Faculties. By ARTHUR NICOLS, F.G.S., F.R.G.S. Illustrated. *In cloth gilt, price 5s., by post 5s. 4d.*

NEEDLEWORK, DICTIONARY OF. An Encyclopædia of Artistic, Plain, and Fancy Needlework; Plain, practical, complete, and magnificently Illustrated. By S. F. A. CAULFEILD and B. C. SAWARD. Accepted by H.M. the Queen, H.R.H. the Princess of Wales, H.R.H. the Duchess of Edinburgh, H.R.H. the Duchess of Connaught, and H.R.H. the Duchess of Albany. Dedicated by special permission to H.R.H. Princess Louise, Marchioness of Lorne. *In demy 4to, 528pp.*, 829 *Illustrations, extra cloth gilt, plain edges, cushioned bevelled boards, price 21s., by post 22s.; with COLOURED PLATES, elegant satin brocade cloth binding, and coloured edges, 31s. 6d., by post 32s. 6d.*

ORCHIDS: Their Culture and Management, with Descriptions of all the Kinds in General Cultivation. Illustrated by Coloured Plates and Engravings. By W. WATSON, Assistant-Curator, Royal Botanic Gardens, Kew; Assisted by W. BEAN, Foreman, Royal Gardens, Kew. Second Edition, Revised and with Extra Plates. *In cloth gilt and gilt edges, price £1 1s., by post £1 2s.*

PAINTING, DECORATIVE. A practical Handbook on Painting and Etching upon Textiles, Pottery, Porcelain, Paper, Vellum, Leather, Glass, Wood, Stone, Metals, and Plaster, for the Decoration of our Homes. By B. C. SAWARD. *In cloth, price 5s., by post 5s. 4d.*

PARCEL POST DISPATCH BOOK (registered). An invaluable book for all who send parcels by post. Provides Address Labels, Certificate of Posting, and Record of Parcels Dispatched. By the use of this book parcels are insured against loss or damage to the extent of £2. Authorized by the Post Office. *Price 1s., by post 1s. 2d.*, for 100 *parcels; larger sizes if required.*

PARROT, THE GREY, and How to Treat it. By W. T. GREENE, M.D., M.A., F.Z.S., &c. *Price 1s., by post 1s. 2d.*

PARROTS, THE SPEAKING. The Art of Keeping and Breeding the principal Talking Parrots in Confinement. By DR. KARL RUSS. Illustrated with COLOURED PLATES and Engravings. *In cloth gilt, price 5s., by post 5s. 4d.*

PATIENCE, GAMES OF, for one or more Players. How to Play 106 different Games of Patience. By MISS WHITMORE JONES. Illustrated. Series I., 39 games; Series II., 34 games; Series III., 33 games. *Each 1s., by post 1s. 2d. The three bound together in cloth, price 3s., by post 3s. 10d.*

PEN PICTURES, and How to Draw Them. A Practical Handbook on the various Methods of Illustrating in Black and White for "Process" Engraving, with numerous Designs, Diagrams, and Sketches. By ERIC MEADE. *In cloth gilt, price 2s. 6d., by post 2s. 9d.*

PERSPECTIVE, THE ESSENTIALS OF. With numerous Illustrations drawn by the Author. By L. W. MILLER, Principal of the School of Industrial Art of the Pennsylvania Museum, Philadelphia. This book is such a manual as has long been desired for the guidance of art students and for self-instruction. The instructions are clearly set forth, and the principles are vividly enforced by a large number of attractive drawings. *Price 6s. 6d., by post 6s. 10d.*

PHEASANT-KEEPING FOR AMATEURS. A Practical Handbook on the Breeding, Rearing, and General Management of Fancy Pheasants in Confinement. By GEO. HORNE. Fully Illustrated. *In cloth gilt, price 3s. 6d., by post 3s. 9d.*

170, Strand, London, W.C. 9

PHOTOGRAPHY (MODERN) FOR AMATEURS. New and Revised Edition. By J. EATON FEARN. *In paper, price* 1s., *by post* 1s. 2d.

PIANOFORTES, TUNING AND REPAIRING. The Amateur's Guide to the Practical Management of a Piano without the intervention of a Professional. By CHARLES BABBINGTON. *In paper, price* 6d., *by post* 6½d.

PICTURE-FRAME MAKING FOR AMATEURS. Being Practical Instructions in the Making of various kinds of Frames for Paintings, Drawings, Photographs, and Engravings. Illustrated. By the REV. J. LUKIN. *Cheap Edition, in paper, price* 1s., *by post* 1s. 2d.

PIG, BOOK OF THE. The Selection, Breeding, Feeding, and Management of the Pig; the Treatment of its Diseases; the Curing and Preserving of Hams, Bacon, and other Pork Foods; and other information appertaining to Pork Farming. By PROFESSOR JAMES LONG. Fully Illustrated with Portraits of Prize Pigs, Plans of Model Piggeries, &c. *In cloth gilt, price* 10s. 6d., *by post* 11s. 1d.

PIG-KEEPING, PRACTICAL: A Manual for Amateurs, based on Personal Experience in Breeding, Feeding, and Fattening; also in Buying and Selling Pigs at Market Prices. By R. D. GARRATT. *In paper, price* 1s., *by post* 1s. 2d.

PIGEONS, FANCY. Containing Full Directions for the Breeding and Management of Fancy Pigeons, and Descriptions of every known Variety, together with all other information of interest or use to Pigeon Fanciers. Third Edition, bringing the subject down to the present time. 18 COLOURED PLATES, and 22 other full-page Illustrations. By J. C. LYELL. *In cloth gilt, price* 10s. 6d., *by post* 10s. 10d.

PIGEON-KEEPING FOR AMATEURS. A complete Guide to the Amateur Breeder of Domestic and Fancy Pigeons. By J. C. LYELL. Illustrated. *In cloth, price* 2s. 6d., *by post* 2s. 9d.

POKER BOOK, THE. How to Play Poker with Success. By R. GUERNDALE. *In paper, price* 1s., *by post* 1s. 2d.

POLISHES AND STAINS FOR WOODS: A Complete Guide to Polishing Woodwork, with Directions for Staining, and Full Information for making the Stains, Polishes, &c., in the simplest and most satisfactory manner. By DAVID DENNING. *In paper, price* 1s., *by post* 1s. 2d.

POOL, GAMES OF. Describing Various English and American Pool Games, and giving the Rules in full. Illustrated. *In paper, price* 1s., *by post* 1s. 2d.

POTTERY AND PORCELAIN, ENGLISH. A Guide for Collectors. Handsomely Illustrated with Engravings of Specimen Pieces and the Marks used by the different Makers. New Edition, Revised and Enlarged. By the REV. E. A. DOWNMAN. [*In the Press.*

POULTRY-KEEPING, POPULAR. A Practical and Complete Guide to Breeding and Keeping Poultry for Eggs or for the Table. By F. A. MACKENZIE. Illustrated. *In paper, price* 1s., *by post* 1s. 2d.

POULTRY AND PIGEON DISEASES: Their Causes, Symptoms, and Treatment. A Practical Manual for all Fanciers. By QUINTIN CRAIG and JAMES LYELL. *In paper, price* 1s., *by post* 1s. 2d.

POULTRY FOR PRIZES AND PROFIT. Contains: Breeding Poultry for Prizes, Exhibition Poultry and Management of the Poultry Yard. Handsomely Illustrated. Second Edition. By PROF. JAMES LONG. *In cloth gilt, price* 2s. 6d., *by post* 2s. 9d.

Published by L. UPCOTT GILL,

RABBIT, BOOK OF THE. A Complete Work on Breeding and Rearing all Varieties of Fancy Rabbits, giving their History, Variations, Uses, Points, Selection, Mating, Management, &c., &c. SECOND EDITION. Edited by KEMPSTER W. KNIGHT. Illustrated with Coloured and other Plates. *In cloth gilt, price* 10s. 6d., *by post* 11s.

RABBITS, DISEASES OF: Their Causes, Symptoms, and Cure. With a Chapter on THE DISEASES OF CAVIES. Reprinted from "The Book of the Rabbit" and "The Guinea Pig for Food, Fur, and Fancy." *In paper, price* 1s., *by post* 1s. 2d.

RABBIT-FARMING, PROFITABLE. A Practical Manual, showing how Hutch Rabbit-farming in the Open can be made to Pay Well. By MAJOR G. F. MORANT. *In paper, price* 1s., *by post* 1s. 2d.

RABBITS FOR PRIZES AND PROFIT. The Proper Management of Fancy Rabbits in Health and Disease, for Pets or the Market, and Descriptions of every known Variety, with Instructions for Breeding Good Specimens. Illustrated. By CHARLES RAYSON. *In cloth gilt, price* 2s. 6d., *by post* 2s. 9d. Also in Sections, as follows:—

General Management of Rabbits. Including Hutches, Breeding, Feeding, Diseases and their Treatment, Rabbit Courts, &c. Fully Illustrated. *In paper, price* 1s., *by post* 1s. 2d.

Exhibition Rabbits. Being descriptions of all Varieties of Fancy Rabbits, their Points of Excellence, and how to obtain them. Illustrated. *In paper, price* 1s., *by post* 1s. 2d.

REPOUSSÉ WORK FOR AMATEURS: Being the Art of Ornamenting Thin Metal with Raised Figures. By L. L. HASLOPE. Illustrated. *In cloth gilt, price* 2s. 6d., *by post* 2s. 9d.

ROSES FOR AMATEURS. A Practical Guide to the Selection and Cultivation of the best Roses. Illustrated. By the REV. J. HONYWOOD D'OMBRAIN, Hon. Sec. of the Nat. Rose Soc. *In paper, price* 1s., *by post* 1s. 2d.

SAILING GUIDE TO THE SOLENT AND POOLE HARBOUR, with Practical Hints as to Living and Cooking on, and Working a Small Yacht. By LIEUT.-COLONEL T. G. CUTHELL. Illustrated with Coloured Charts. *In cloth, price* 2s. 6d., *by post* 2s. 9d.

SAILING TOURS. The Yachtman's Guide to the Cruising Waters of the English and Adjacent Coasts. With Descriptions of every Creek, Harbour, and Roadstead on the Course. With numerous Charts printed in Colours, showing Deep water, Shoals, and Sands exposed at low water, with sounding. *In Crown* 8vo., *cloth gilt*. By FRANK COWPER, B.A.

Vol. I., the Coasts of Essex and Suffolk, from the Thames to Aldborough. Six Coloured Charts. *Price* 5s., *by post* 5s. 3d.

Vol. II. The South Coast, from the Thames to the Scilly Islands, with twenty-five Charts printed in Colours. *Price* 7s. 6d., *by post* 7s. 10d.

Vol. III. The Coast of Brittany: Descriptions of every Creek, Harbour, and Roadstead from L'Abervrach to St. Nazaire, and an Account of the Loire. With twelve Charts, printed in Colours. *Price* 7s. 6d., *by post* 7s. 10d.

Vol. IV. The West Coast, from Land's End to Mull of Galloway, including the East Coast of Ireland. With thirty Charts, printed in Colours. *Price* 10s. 6d., *by post* 11s.

ST. BERNARD, THE. Its History, Points, Breeding, and Rearing. By HUGH DALZIEL. Illustrated with Coloured Frontispiece and Plates. *In cloth, price* 2s. 6d., *by post* 2s. 9d.

ST. BERNARD STUD BOOK. Edited by HUGH DALZIEL. *Price* 3s. 6d. *each., by post* 3s. 9d. *each.*

170, Strand, London, W.C. 11

Vol. I. Pedigrees of 1278 of the best known Dogs, traced to their most remote known ancestors, Show Record, &c.
Vol. II. Pedigrees of 564 Dogs, Show Record, &c.
SEA-FISHING FOR AMATEURS. Practical Instructions to Visitors at Seaside Places for Catching Sea-Fish from Pier-heads, Shore, or Boats, principally by means of Hand Lines, with a very useful List of Fishing Stations, the Fish to be caught there, and the Best Seasons. By FRANK HUDSON. Illustrated. *In paper, price* 1s., *by post* 1s. 2d.
SEA-FISHING ON THE ENGLISH COAST. The Art of Making and Using Sea-Tackle, with a full account of the methods in vogue during each month of the year, and a Detailed Guide for Sea-Fishermen to all the most Popular Watering Places on the English Coast. By F. G. AFLALO. Illustrated. *In cloth, price* 2s. 6d., *by post* 2s. 9d.
SEASIDE WATERING PLACES. A Description of the Holiday Resorts on the Coasts of England and Wales, the Channel Islands, and the Isle of Man, giving full particulars of them and their attractions, and all information likely to assist persons in selecting places in which to spend their Holidays according to their individual tastes. Illustrated. Seventh Edition. *In cloth, price* 2s. 6d., *by post* 2s. 10d.
SHADOW ENTERTAINMENTS, and How to Work Them: being Something about Shadows, and the way to make them Profitable and Funny. By A. PATTERSON. *In paper, price* 1s., *by post* 1s. 2d.
SHAVE, AN EASY: The Mysteries, Secrets, and Whole Art of, laid bare for 1s., *by post* 1s. 2d. Edited by JOSEPH MORTON.
SHEET METAL, WORKING IN: Being Practical Instructions for Making and Mending Small Articles in Tin, Copper, Iron, Zinc, and Brass. Illustrated. Third Edition. By the Rev. J. LUKIN, B.A. *In paper, price* 1s., *by post* 1s. 1d.
SHORTHAND, ON GURNEY'S SYSTEM (IMPROVED), LESSONS IN: Being Instructions in the Art of Shorthand Writing as used in the Service of the two Houses of Parliament. By R. E. MILLER. *In paper, price* 1s., *by post* 1s. 2d.
SHORTHAND, EXERCISES IN, for Daily Half Hours, on a Newly-devised and Simple Method, free from the Labour of Learning. Illustrated. Being Part II. of "Lessons in Shorthand on Gurney's System (Improved)." By R. E. MILLER. *In paper, price* 9d., *by post* 10d.
SHORTHAND SYSTEMS; WHICH IS THE BEST? Being a Discussion, by various Experts, on the Merits and Demerits of all the principal Systems, with Illustrative Examples. Edited by THOMAS ANDERSON. *In paper, price* 1s., *by post* 1s. 2d.
SKATING CARDS: An Easy Method of Learning Figure Skating, as the Cards *can be used on the Ice. In cloth case,* 2s. 6d., *by post* 2s. 9d.; *leather,* 3s. 6d., *by post* 3s. 9d. A cheap form is issued printed on paper and made up as a small book, 1s., *by post* 1s. 1d.
SLEIGHT OF HAND. A Practical Manual of Legerdemain for Amateurs and Others. New Edition, Revised and Enlarged. Profusely Illustrated. By E. SACHS. *In cloth gilt, price* 6s. 6d., *by post* 6s. 10d.
TAXIDERMY, PRACTICAL. A Manual of Instruction to the Amateur in Collecting, Preserving, and Setting-up Natural History Specimens of all kinds. With Examples and Working Diagrams. By MONTAGU BROWNE, F.Z.S., Curator of Leicester Museum. Second Edition. *In cloth gilt, price* 7s. 6d., *by post* 7s. 10d.
THAMES GUIDE BOOK. From Lechlade to Richmond. For Boating Men, Anglers, Picnic Parties, and all Pleasure-seekers on the River. Arranged on an entirely new plan. Second Edition, profusely illustrated. *In paper, price* 1s. *by post* 1s. 3d.; *cloth,* 1s. 6d., *by post* 1s. 9d.

TOMATO AND FRUIT GROWING as an Industry for Women. Lectures given at the Forestry Exhibition, Earl's Court, during July and August, 1893. By GRACE HARRIMAN, Practical Fruit Grower and County Council Lecturer. *In paper, price 1s., by post 1s. 1d.*

TOMATO CULTURE FOR AMATEURS. A Practical and very Complete Manual on the Subject. By B. C. RAVENSCROFT. Illustrated. *In paper, price 1s., by post 1s. 3d.*

TRAPPING, PRACTICAL: Being some Papers on Traps and Trapping for Vermin, with a Chapter on General Bird Trapping and Snaring. By W. CARNEGIE. *In paper, price 1s., by post 1s. 2d.*

TURNING FOR AMATEURS: Being Descriptions of the Lathe and its Attachments and Tools, with Minute Instructions for their Effective Use on Wood, Metal, Ivory, and other Materials. Second Edition, Revised and Enlarged. By JAMES LUKIN, B.A. Illustrated with 144 Engravings. *In cloth gilt, price 2s. 6d., by post 2s. 9d.*

TURNING LATHES. A Manual for Technical Schools and Apprentices. A guide to Turning, Screw-cutting, Metal-spinning, &c. Edited by JAMES LUKIN, B.A. Third Edition. With 194 Illustrations. *In cloth gilt, price 3s., by post 3s. 3d.*

VAMPING. A Practical Guide to the Accompaniment of Songs by the Unskilled Musician. With Examples. *In paper, price 9d., by post 10d.*

VEGETABLE CULTURE FOR AMATEURS. Containing Concise Directions for the Cultivation of Vegetables in Small Gardens so as to insure Good Crops. With Lists of the Best Varieties of each Sort. By W. J. MAY. Illustrated. *In paper, price 1s., by post 1s. 2d.*

VENTRILOQUISM, PRACTICAL. A thoroughly reliable Guide to the Art of Voice Throwing and Vocal Mimicry, Vocal Instrumentation, Ventriloquial Figures, Entertaining, &c. By ROBERT GANTHONY. Numerous Illustrations. *In cloth, price 2s. 6d., by post 2s. 9d.*

VIOLINS (OLD) AND THEIR MAKERS: Including some References to those of Modern Times. By JAMES M. FLEMING. Illustrated with Facsimiles of Tickets, Sound-Holes, &c. Reprinted by Subscription. *In cloth, price 6s. 6d., by post 6s. 10d.*

VIOLIN SCHOOL, PRACTICAL, for Home Students. Instructions and Exercises in Violin Playing, for the use of Amateurs, Self-learners, Teachers, and others. With a supplement on "Easy Legato Studies for the Violin." By J. M. FLEMING. *Demy 4to, price 9s. 6d., by post 10s. 4d. Without Supplement, price 7s. 6d., by post 8s. 1d.*

WAR MEDALS AND DECORATIONS. A Manual for Collectors, with some account of Civil Rewards for Valour. Beautifully Illustrated. By D. HASTINGS IRWIN. *In cloth, price 7s. 6d., by post 7s. 10d.*

WHIPPET AND RACE-DOG, THE: How to Breed, Rear, Train, Race, and Exhibit the Whippet, the Management of Race Meetings, and Original Plans of Courses. By FREEMAN LLOYD. *In cloth gilt, price 3s. 6d., by post 3s. 10d.*

WILDFOWLING, PRACTICAL: A Book on Wildfowl and Wildfowl Shooting. By HY. SHARP. *Demy 8vo, price 7s. 6d.* [*In the Press.*

WIRE AND SHEET GAUGES OF THE WORLD. Compared and Compiled by C. A. B. PFEILSCHMIDT, of Sheffield. *In paper, price 1s., by post 1s. 1d.*

WOOD CARVING FOR AMATEURS. Full Instructions for producing all the different varieties of Carvings. 2nd Edition. Edited by D. DENNING. *Price 1s., by post 1s. 2d.*

EXTRA
Special Supplements

ARE GIVEN **FREE** WITH

The Bazaar, Exchange & Mart.

DURING THE SEASON.

Each Supplement is for a special body of Readers, as follows:

THE PHILATELISTS' SUPPLEMENT,
THE LADIES' SUPPLEMENT,
THE DOG OWNERS' SUPPLEMENT,
THE PHOTOGRAPHERS' SUPPLEMENT,
THE CYCLISTS' SUPPLEMENT,
THE ANGLERS' SUPPLEMENT,
THE MUSICIANS' SUPPLEMENT,
THE GARDENERS' SUPPLEMENT,
THE SHOOTERS' SUPPLEMENT,

And each is Splendidly Illustrated.

WITH
EXTRA SUPPLEMENTS
FREE.

The Bazaar,
Exchange and Mart,
and
Journal of the Household.
Published Every Monday, Wednesday, and Friday.

ILLUSTRATED.] ESTABLISHED 26 YEARS. [PRICE TWOPENCE

Cadbury's Cocoa
ABSOLUTELY PURE
No Che...
Extra

NEWHAM'S LINCOLNSHIRE FEATHER BEDS.

KNIVES, FORKS, SPOONS.

PRIVATE SALES and BARGAINS
For Every Description of Property by PRIVATE PERSONS, there is no medium equal to this Paper.
Price 2d.
OF ALL NEWSAGENTS.

DRESS F...

PATT...
AUTUMN &
TROUSERS
(Made to Measure)
21/- CHESTERFI...
OVERC...
THE OLD WOOLLEN MILLS CO.,
7, Park Lane, Leeds.

Dinneford's Magnesia
For ACIDITY of the STOMACH, HEARTBURN, GOUT, and HEADACHE, INDIGESTION

Benger's Food.
For INFANTS, CHILDREN, INVALIDS, and THE AGED.

HOW TO MAKE
A CAMERA, A HAND CAMERA, A LANTERN, A TRIPOD.
"The Cyclopedia of Photo-Brasswork."

[WITH SUPPLEMENTS.]

OFFICE: 170, STRAND, LONDON, W.C.

BARGAINS

IN ALL KINDS OF

Private Property

of every description, are readily obtained through

The Bazaar,
The Exchange and Mart

(Established 27 Years).

THE BEST MARKET IN THE WORLD FOR THE DISPOSAL OF SURPLUS PROPERTY.

Lots of Interesting, Useful, and Valuable Information on a vast number of popular subjects, with Numerous Illustrations.

THE MOST USEFUL PAPER IN LONDON.

GET A COPY AND SEE.

Price **2d.**, at all *Newsagents* and *Railway Bookstalls.*

Offices: 170, Strand, London, W.C.

The Diet of all Lucky Dogs

SPRATTS PATENT
DOG CAKES.

Pamphlet on CANINE DISEASES GRATIS.

SPRATTS PATENT LIMITED, BERMONDSEY, S.E.

www.ingramcontent.com/pod-product-compliance
Lightning Source LLC
Chambersburg PA
CBHW022126160426
43197CB00009B/1172